Wild!

The True Nature of Animal Trafficking

Jérôme Pensu
with Bertrand Ferrier

Wild!

The True Nature of Animal Trafficking

Max Milo

www.maxmilo.com
ISBN: 978-2-31502-277-9

To my parents

Foreword

When you think of "wildlife trafficking", you'll think of the clichés that have dominated the media: smuggling in rare exotic animals to enrich a few smugglers and sponsors, crates opened by customs at Orly, snow leopard skins seized at the border and so on. The reality is quite different, or at least more complex and far more diverse.

There is an ordinary, less visible, more chronic form of trafficking that goes under the radar. The recreational wildlife industry has generated a world that is unsuspected by the layman, who only has access to the glitz and glamour of the shows, a world with its big groups, its protectors, its suppliers, its freedmen, its little hands, its cleaners, its active and passive corruption.

Jérôme Pensu, with the efficiency and doggedness of an ant on steroids, encyclopedic regulatory knowledge and a solid grounding in the biology of species, brings to light, demonstrates and dismantles this universe worthy of Mario Puzo and Ettore Scola, and is not content with imprecations and suppositions. In fact, every assertion is substantiated, every accusation supported by written proof. It took an unimaginable amount of perseverance and self-sacrifice to gather and publish these facts, against all odds. Involved in the protection of nature and wildlife from an early age, he has always been driven by passion and exemplary ethics. I got to know him in the late 1990s, and helped him set up a seabird removal center in the Landes region of France during the *Prestige* oil spill in 2003. At the time, we had to kick against the anthills of inertia, favoritism, conflict of interest and POLMAR fund-grabbing to succeed in setting up this de-oiling unit, whose results in terms of releases and long-term survival are unrivalled in France to this day. This first common battle, a shared vision of wildlife conservation and the same intolerance of injustice forged a solid friendship and taught me to get to know Jérôme.

Rather than write a preface that merely contextualizes the work that follows, I wish above all to do justice to the man whose commitment, uprightness, passion, ethics and professionalism have always been constant. It is these very qualities that have, over the years, provoked against him the outburst of those who, deprived of these qualities and having moreover hidden and illicit interests to defend, have endeavoured to hinder his action, to tarnish his reputation, so true is it that honesty exasperates the dishonest. Having witnessed a large part of the war waged against him, I was able to gauge, in the face of the resources deployed, the scale of the financial stakes and the extent to which some were compromised.

Defamation, abusive accusations, police custody, invective, blacklisting, administrative bullying... others like Jérôme would have given up, faced with the high wall, and preferred a little more comfort and tranquillity.

And then... just as everything seemed closed and played out, a tipping point seems to occur. Such constancy of action raises questions, provokes emulation. The lines are moving: courageous civil servants and magistrates are refusing to turn a blind eye. The public consensus on captive fauna is crumbling, tongues are wagging, animal malaise in many establishments is being revealed and decried, and the ecological emergency and the vital need to act visibly and effectively for biodiversity are asserting themselves. As a result, the rhetoric and habits of the old world, the colonial vision of zoos recently cloaked in the mantle of conservation, are losing credibility and becoming obsolete in the face of today's challenges. Nevertheless, like any system in its death throes, this world of lucrative and recreational captivity is having dangerous convulsions, as witnessed by the decree of October 8, 2018, which liberalizes and facilitates the keeping of wild animals, extremely revealing of the interconnections between certain high circles of the administration and the captivity *lobby*.

After Jérôme's observation, the remedy: the rewilding of trafficked wildlife, the priority return of trafficked wildlife to the wild and, if that's impossible, to sanctuary. And to get animals out of the vicious circle of animal exploitation which, until recently, was the only alternative. The simple premise of releasing captive wildlife into the wild, though based on numerous successes and in-depth knowledge, is being challenged by the proponents of captivity, because it shakes the foundations of the captivity industry and exposes the sham of the conservation alibi and animal ambassadors, enabling public awareness and progress in protection.

This book should become bedside reading for any official in charge of control and for anyone sensitive to the animal cause, species conservation and the degradation of biodiversity. In any case, let's hope it provides a salutary jolt!

Gilles Moyne
Capability Director, Centre Athénas

Disclaimer: In order to preserve the privacy of those involved, some names have been changed. The essence of the events remains faithful to reality as I experienced it.

Prelude

Amnéville Zoo. 15 h 30. I've just dropped off three birds of prey from the Alca Torda care center I designed and run. These birds, suffering from several pathologies, are unable to survive in the wild. I had two options: euthanasia or placement in a zoo. I opted for the latter. This will be the last time I make such a mistake.

But I didn't choose Amnéville Zoo by chance. I've had a fairly long-standing relationship with Michel Louis, its director. I like his outspokenness and unconventionality, and I have great confidence in him, until proven otherwise. A rarity among his peers, Michel accepts that he is not a savior of wildlife. He describes himself as an animal presenter. However, he claims to do it to the best of his ability, with a sincere love of wildlife; in exchange for which, every year, he claims to sign a cheque for €500,000 in favour of conservation. Among other things, he has funded Kalaweit, one of the major *in situ* protection and conservation projects for Gibbons and Siamangs, for many years.

I like this guy, with whom I can both disagree and agree, without these differences leading to trench warfare. I'm not fooled by the consequences of *business* on animals, but I like the fact that this zoo director doesn't use the pretext of conservation to justify his business.

After dropping off the animals in the aviary, I meet Michel in his office. Coffee, pleasantries, a cigarette, an exchange of news—all goes well. Until, from his desk drawer, the man pulls out a book of cheques. Puzzled, I ask him what he's up to.

—Well, I'm writing you a check for the birds! he blurts out in a tone of obviousness.

My eyes round, I stammer:

—But... but that's out of the question!

—What's wrong with that?

—Michel, I've just given you some wild animals.

—I know you did.

—Animals that come from the wild.

—What about them?

—They're not for sale!

—No, okay, I won't buy them from you, but you and I know how much they're worth.

—Uh... not me, no.

—Well, at the very least, a griffon vulture and two short-toed eagles are worth about fifteen thousand euros.

I understand that, for a trader, a wild animal has a market value, but I don't hesitate for a second:

—Michel, thank you, but you can keep your fifteen thousand euros.

—But why?" he insists.

—Just imagine! I'm arriving tomorrow with new specimens that aren't really impregnated, but which I'm telling you are. I'm going to pocket my fifteen grand again?

—Of course you will.

—At what point do you think we don't overflow?

His eyes open like saucers. The guy's a shopkeeper, not a crook. He just crashed.

—Damn it, Jérôme, are you telling me that's how it works?

—Of course I'm telling you! You think it's easy to resist the call of fifteen thousand euros? Even if the guy's honest, he may accept it for very good reasons, like rebuilding an aviary or saving a center in trouble. Nevertheless, it's illegal sales—and therefore trafficking—that we're talking about…

Michel's astonishment tells me that either he deserves an Oscar for Best Actor, or that no-one has ever said that to him. And with good reason! In an ideal, honest world, when I have non-releasable animals that I have to euthanize, I could take them to a zoo which, knowing that it will benefit from them, would pay me back or make a donation to my structure. The transaction wouldn't shock anyone. In reality, this type of exchange turns wildlife into a commodity and allows protected animals to be held in captivity, where they cannot be captured and sold.

This is the *business model* that unites many care centers and zoos. It's a crime. It's called trafficking, i.e. an illegal, or disguised, commercial transaction involving animal species whose capture for the purposes of trade is prohibited.

Wildlife trafficking, the illegal offshoot of a legal trade, relies on a captivity industry that is always on the lookout for animals to feed its *business*. Although it receives less media coverage than drug dealing or arms sales, wildlife trafficking is one of the most profitable businesses in France. It represents several tens of millions of euros every year. Far from hiding, the traffickers are well established. These are exotic bird farms, zoos, and sometimes even care centers and NGOs that profit from what they are supposed to be fighting against. Their best shields?

- The French administration;
- *lobbies*;
- and lies.

I'll absolve the *lobbies* of all blame. Their role is to plead in favor of an industry that is fatal both to animals and to the planet, and therefore to the human species. The problem is that they find interlocutors who are quick to be persuaded... notably in the administration, which long ago took over the reins of the country. And what about politicians? Supposedly responsible for our country's ecological health, they aren't even complicit in the situation—it's worse: with rare exceptions, they're unaware of the preponderant role they're playing in this drama. With their eyes closed, they sign legislation, allow by-laws to be published, take part in symposia and organize conferences. In other words, they do what their administration tells them to do. In this way, they eradicate priceless species while flouting international laws. By swallowing and regurgitating misleading speeches aimed at the general public, by strutting around in big zoos in front of the cameras, they encourage the importation and commodification of wildlife, under the impression that they are pursuing a laudable environmental policy and fulfilling their mission as political decision-makers in the service of the public good.

*

For me, writing a book on this subject has never been a goal. But how can I avoid it? Time and again, I've tried to draw the attention of journalists to the scale of the trafficking and the involvement of government officials in the scandal. I met with numerous journalists, and presented my observations at length, without succeeding in shaking the coconut tree as I had hoped. I didn't understand why, until Laure Noualhat, then a journalist with *Libération,* said to me: "Your story can't fit in a paper. It can't even fit in a special report. Why don't you write a book?"

Writing doesn't make me fantasize or tremble. I've written numerous reports, studies and expert opinions. Never a book. Today, I've decided to get down to it, because I feel I have no choice if I want to:

- remain coherent with myself ;
- alert politicians, who are guilty if not responsible;
- denounce the ruthless methods of the traffickers and the dramatic consequences of their *business*;
- and to say to every reader: "Maybe you're a customer of pet shops that sell exotic animals. Maybe you even have wild animals at home, and think they're lucky to be under your protection. Perhaps you visit zoos. Don't worry about criticism or moralizing—this isn't the project! Simply, if you recognize yourself or someone close to you in these portraits, dare to read the following pages. You'll understand what appalling trafficking these seemingly innocent choices are".

This book comes from this feeling of need and duty to inform. It also comes from the urgency of the current situation. I can no longer keep quiet about my anger, the anger of a man who has devoted his life to the protection of wildlife and who, every day, watches almost helplessly as those who have turned it into a lucrative *business* go unpunished. Based on the principle that he who knows and remains silent is an accomplice, I decided to tell the story.

The aim is neither to fire in all directions nor to tar everyone with the same brush, at the risk of fuelling the "everyone's rotten" sentiment. My anger does not rob me of my lucidity or my objectivity. I'm neither embittered nor suffering from ecolodogmatic fanaticism. On the other hand, the revolt that drives me and motivates me to write is rooted in three ideas:

- traffickers, large and small, are damaging and even destroying our common heritage;
- thanks to the incredible power of misinformation, these traffickers are making millions of French people complicit in the destruction of ecosystems and biodiversity;
- yet the consequences of this exploitation have dramatic repercussions for mankind, which is directly dependent

on the balance of ecosystems seriously endangered by trafficking.

My findings, which I set out in this open-hearted book with incontrovertible documents and evidence, are astounding. For thirty years, regulations have been undermined by high-ranking civil servants to the benefit of captive industry *lobbies*. The environmental police have been vitiated by the same *lobbies* that defend breeders, importers and zoo owners; control of the animal captivity industry is corrupted by governmental unwillingness, which is illustrated and even fuelled by a lack of resources and skills; penalties are virtually non-existent or out of all proportion to the offence, with the result that the last frameworks in force are ineffective.

In this book, I tell the story of how, against all odds, a bird-mad kid became an expert in the fight against wildlife trafficking. I'll explain how, because I wanted to defend a cause and stood up to traffickers, I went from being a (sometimes judicial) expert in "captive wildlife", a member of the commission of the same name and a trainer for the environmental police, to being a police officer registered on the TAJ (tableau des antécédents judiciaires) in less time than it takes to say it.

Today, I am both saddened and dismayed by the catastrophic situation in which we find ourselves.

I'm appalled by the quantitative and structural scale of drug trafficking in France, which is largely fuelled by the complicity of the administration and certain associations, and totally ignored by the political authorities.

But I'm also frightened by the disappearance of animal species and the destruction of ecosystems.

That's why, on the basis of these subjective experiences and objective observations, I tell the story of how it is still possible to change the course of history, even if it will be very complicated, and even if sometimes, faced with the scale of the problem, I believe in it less and less.

In reality, this story isn't just about me.

The gravity of the situation must stimulate our desire for change. As for me, every morning I'm as determined and determined as on my first day. I'll never give up, and I'd like to think I'm not the only one.

In short, with this book, I want to help you understand:

- why we are all partly responsible for the state of our planet, for ourselves and for future generations ("we don't inherit the world from our parents, we borrow it from our children", according to a quote sometimes attributed to Antoine de Saint-Exupéry);
- how certain mechanisms and interests weaken it;
- and, above all, how each and every one of us can take concrete action, even if it's only on an individual scale and as a citizen, to refuse the inevitable disappearance—ever slower and ever more dramatic—of wildlife, in order to successfully contribute to its preservation.

I also hope that this book will make a contribution and serve as proof that those who have committed all the misdeeds described here will be removed from their posts and that the mafia that enables trafficking via the Ministry of Ecology will be dismantled, and that these players will be unmasked, otherwise... It will have been for nothing!

1.
Chicks in the Drawer

Sometimes, destinies are self-evident. Some become doctors, gendarmes or workers like dad, granddad and great-granddad; others take up the torch of a farming or industrial business to perpetuate a family capital, tradition and way of life. Whether self-fulfilling or inevitable, these orientations seem to impose themselves. On the other hand, it wasn't written in stone that I would dedicate my life to wildlife—far from it! I wasn't born on the savannah, my parents didn't look after hippos and, I must confess, I wasn't taken in by wolves after being abandoned in the forest one winter's evening. On the other hand, I was greatly influenced by animated series such as *The Mysterious Cities of Gold, Heidi, Daktari* and *Goldorack*, where wildlife, wide open spaces, justice and vigilantes undoubtedly guided my choices and my taste for everything to do with nature, animals and respect for the law.

For this book, I tried with great difficulty to pinpoint the moment when I realized which road I had to take. Everything changed the day my path crossed that of an injured seagull.

*

Summer 1984. I'm fifteen years old.

Around five o'clock, after a day on the beach in the sun of the South of France, with the two cousins of my girlfriend at the time, we spot a seagull at the water's edge. Its behavior suggested that it was in distress. He had a wing hanging down and a fishing line sticking out of his beak.

Any kid my age probably wouldn't have paid any attention to the scene, but faced with an injured animal, it's impossible for me to look away. After a strategic flap, we succeed—not without difficulty—in catching the individual. On first examination of the uncooperative gull, I instinctively decide to immobilize the wing with a bandage, so that it can regain normal carriage. I have no idea of the medical consequences of this immobilization, but decide to keep the animal quiet in the garage for 24 hours, while I come up with a solution for the hook. That's my main concern. I'm not a vet, so I imagine the hook is stuck deep in the animal's viscera. I spend an almost sleepless night looking for *the* right idea. As the first rays of sunshine filter through the shutters of my bedroom, I make my decision: it's time to operate.

At breakfast, I announce my decision. My cousins are skeptical. Yet my plan of attack is clear. The neighbor will supply us with the ether needed for anesthesia. Using my DIY skills, I improvise a scalpel with a razor blade cut in half and mounted on a piece of wood. Ether is poured liberally onto a cloth, which we wrap around the animal's beak until it falls asleep. A simple pull on the fishing line reveals the throat's incredible unfolding capacity. In fact, it makes sense, since the gull is capable of swallowing large prey. The bird's stretched throat reveals the hook simply planted under a thin skin, right next to the windpipe. The situation isn't as bad as I'd imagined.

All of a sudden, the pressure I had put on myself disappears. I make a very slight incision in the skin on which the hook is hooked. With the utmost delicacy, I managed to extract the piece of metal. Half an hour later, the bird opens its eyes. We'll keep it safe for the duration of our vacation. It will become the house mascot. At the end of our stay, his wing is restored. We watch him fly away and disappear. Our success fills us with joy. How good it is to feel useful! Clearly, at that moment, I said to myself: "This is what I want to do with my life, this is where I belong". Of course, nothing is set in stone. I'd love to become a vet, but I have an absolute aversion to anything academic. So what course of study could I take to look after wild animals? No idea, but the intuition and desire are there. They'll never leave me.

*

This key event didn't happen by chance. As far back as I can remember, I've always loved birds. Behind the windows of our suburban bungalow, I could spend hours watching them. At the bottom of the garden, my father helped me build a bird feeder and nesting box, which I use as my observation site. With my childlike eyes, I love observing this winged fauna. I marvel at their colors. I envy their apparent freedom. I appreciate their vivacity. I'm fascinated by something so simple, yet so extraordinary: THEY FLY! I never tire of admiring this phenomenon. With the few ornithological books available in my parents' library, I learn to identify the different species, and learn about their diet and habits. My curiosity became a passion.

As well as European birds, I'm particularly fond of exotic birds—their colourful plumage, the variety of their lifestyles, their incredible diversity—without understanding how animals living so far away can reach us. But it's not just birds that fill me with joy: I love all animals. There's no doubt that one man helped to develop this love: my maternal grandfather. He was a doctor and, by his own admission, regretted having been "too lazy" to become a veterinary surgeon, because in his day, it took a year longer to study for that title than it did to practise medicine. I inevitably enjoyed a very privileged status in his eyes. He had dreamed of having a son, but instead had five daughters whom, as a good country doctor, he helped to bring into the world. When my sister, the first of ten grandchildren, was born, a little voice whispered to him, "Another girl!"

As the first boy in the family, and the first to be born by his own hand, I'm aware that I was a little privileged, until I was dethroned by the other seven. So, on my vacations in the Vendée, he takes me to the farms he visits on his rounds. While he treats his patients, I stay in the yard with the animals. My passion for animals is nourished by these moments when I hang around these modest farms. Through observation and contact with the species I come into contact with, I become immersed in the animal world. With each experience, my fascination grows. Imagine what a cow or a horse can mean to a five-year-old boy, who finds them gigantic and so attractive... On vacations, I refine my knowledge with the books in my grandfather's possession and, years later, with those I buy with the money I earn from washing the cars in our house. I had to wash a lot of cars to perfect my knowledge and fuel my almost obsessive desire to be with animals and help them!

*

Reflecting on the genesis of this passion for birds, I also remember that, every year, my father used to take me to the Salon de l'Agriculture. He runs a stand there on behalf of the Alibert company he works for. It's my breath of fresh air for the year. I look forward to it. The year is 1979. I'm ten years old, and I'm wandering through the aisles of what is still called France's biggest farm. I was in awe of the breeding facilities that, in my naivety at the time, made me dream, notably Phénix Gibier and Ferme de Beaumont. I admired exotic ducks, ostriches, pheasants and parrots for the first time.

At the end of one of these shows, I pull off what is for me a tour de force. In exchange for an Alibert tray, collected from my father's stand, I obtained ten day-old chicks from a manufacturer of heat lamps. After the party, their destiny was undoubtedly death. The achievement lay not in negotiating with the exhibitor, but rather in convincing my father to let me take my pack of chicks home. I'm thrilled. My father less so. He admits he only agreed because he was convinced my protégés would be dead within 24 hours.

Not a chance: I put all my heart into this rescue. First I raise the chicks in my office drawers, then in the garage. Much to my father's dismay, none perish. They all grow up healthy. I'm only allowed to keep two, as they're all males. The neighbors would no doubt take a dim view of a cacophonous awakening caused by this improvised breeding. They remained my pets for a long time. At the time, I imagined that to love an animal was to own it. As I studied wild animals and came to understand them better, I learned that not necessarily. Depending on the species, it can be quite the opposite.

2.
The Law of Nature

A year before I became humanity's most short-lived—but not least effective!—of mankind, I began a career as a caretaker for exotic ducks. In 1983, my parents, my sister, my brother and I were living in the Île-de-France region, in a housing estate of some seven hundred detached houses built on the site of a former wetland. Clearly, such construction would no longer be permitted today, given the environmental importance of the area destroyed to build these homes. To drain the site, the developer created an artificial lake at the center of which sits an island.

The lake is home to wild ducks, and an extraordinary playground for me. Two co-owners decided to complement this natural setting by purchasing exotic ducks with funds allocated by the co-ownership. Native ducks, barnacle geese, harvest geese, widgeon ducks and even Australian black swans now live side by side. Some winters, the lake freezes so hard that the residents of the residence find themselves on the ice and, like an army of Jesus Christ, walk on the water. All that's left for the birds is a tiny patch of running water, where the feeder channel creates enough eddies to keep the water moving vemently.

Between the cold and the lack of food, several individuals seem to be in very bad shape. They climb onto the banks where they remain, beak in plumage and trembling. In their native habitat, these animals feed on grass, aquatic plants and small water gastropods. Here, winter is harsh, so they have nothing to eat. Without understanding the plight of the birds, the locals are delighted to be able, for once, to get close to them, whereas the rest of the time they're content to see them in the distance. The drama of some becomes a distraction for others.

This freezing winter crystallizes my passion for wild birds and my desire to act. My instinct is to intervene, and urgently. I've always been a shy, introverted child. Nevertheless, when it comes to defending a cause, I find the necessary resources to get the job done. So I plucked up the courage to negotiate access to an old boathouse with the condominium owners, and set about capturing birds for safekeeping. It's a first for me. Thanks to my instinct, my sense of observation and my kid-like alertness, the birds are captured, sheltered and fed. They stayed there until the end of the cold snap.

I became, if not famous, then at least identified as "the boy who takes care of the birds". So much so that, some time later, having heard of me, the village fire department came knocking at the house: they needed a guy with a little knowledge of birds to help them free a swan that had been reported to them as trapped in the ice. This would be my first intervention as an assistant to a uniformed authority, but it would also be another learning experience. Instead of a swan trapped in the elements, we're dealing with a bird weakened by winter and just sitting on the ice. I find myself embarked in the firefighters' boat, resting on frozen water and pulled by several of them like Santa's sleigh. I must admit to feeling a certain pride. The fact that those whose uniforms represent authority call on me to assist them gives me a feeling of recognition that heightens my motivation.

However, my knowledge remains limited, and I'm about to get a violent demonstration of this. The swan will be caught without difficulty, but with great caution as to the danger posed by the beak, I thought foolishly. But on the way back to the hangar, my arm, which was encircling the body of the bird in distress, shows signs of weakness given the size and weight of the animal. He takes the opportunity to free his left wing. Surrounded by firefighters, I woke up with a bump on my head. The swan had simply knocked me unconscious with a violent sweep of its wing to free itself from my grasp. The danger with a bird of this size is the wing, not the beak, another lesson I'm not about to forget.

*

The lake becomes my learning ground; the birds are my favorite companions. As soon as school's out, I'm off to check if the ducks are there, if the geese have enough food, if any eggs have hatched. With some accomplices, defying the formal ban on venturing out onto the water, we make our way to the artificial island to build nesting boxes for the ducks, to keep them safe from rats. We feel we're not doing good, but doing what needs to be done.

Not everyone shares our conviction. I was fourteen when I faced my first adversaries and, through them, the thorny issues that would accompany the rest of my life in the service of wildlife. To open the ball of unpleasantness, two women in trent on stage. They're the ones who argued for the acquisition of the exotic ducks on the lake, and they rant and rave as they see me bustling around them. They come to my parents, complain, demand, storm and threaten. According to them, we mustn't interfere. Ducks are old enough to fend for themselves. How do they do it in the wild? They don't need us, they're ducks!

The argument goes, they know nothing about it. Of course local ducks are genetically prepared for the life that awaits them! But when you release exotic animals into a native environment, it's just as obvious that they're not adapted to it. "In the wild, there are two solutions: either these animals live in regions with a sometimes tropical climate, sheltering them from the vagaries of the weather; or they are migratory, precisely to escape the problems of climate and food. When the frost hits, exotic ducks can't survive without a little help, that's how it is; and, whether our neighbors like it or not, I have no intention of letting animals die... especially as they suffer from an additional handicap: they have little chance of migrating, having been mutilated so that we can have the pleasure of keeping them available to contemplate.

It's a sinister habit. The majority of waterbirds—whether in parks, zoos or private homes—are amputated in their first days. When we try to disguise this amputation, we use a much softer, much more hypocritical word than "mutilate" or "amputate": we say that the birds have been spayed. In other words, we cut off "the hand", i.e. the wing tip. In other words, they've been mutilated. Amputation is not inevitable. On a wing, the first primary remiges are used for flying, the others for gliding. Instead of cutting off part of the wing to keep the birds in a given space, we can "remiger" them, i.e. cut off the flight feathers. The exercise is as painless for the animal as a haircut for a human. As the flight feathers grow back, the operation must be repeated twice a year. So, to be on the safe side, if scruples don't stifle you, you cut off the animal's hand, where the feathers of freedom are attached, and you're sure the duck will never leave you again.

*

And so, groping my way, I come face to face with the first elements of a reflection on the ethics of animal care and husbandry. I begin to understand that the man who cares for wild animals should always remember the mantra that, "with great power comes great responsibility". With the ducks in the condominium, I'm developing a reasonable closeness. In that sense, I've tamed them. By bringing them food, they accept my presence. For a kid, this sign of trust from a bird that's easily scared off is a real treat! In my joy, I'm unaware of the dangerous path I'm taking: it's never a good idea to tame a wild animal, made for freedom. Often, the desire to tame an animal is based on a beautiful sentiment that combines affection and the desire to protect it. Unfortunately, it has its dark side. It's also a desire for possession. And nothing feeds trafficking more than the desire to possess.

Traffickers flatter and excite this desire and instinct for possession. But what's even more astonishing is that it's also driving a new generation of eco-animalists. They tolerate captivity on the pretext that a wild animal is in danger in the wild, especially when it is likely to be prey. For them, releasing an animal increases its risk of violent death. This is a fact *and* it's nature, as beautiful as it is dangerous, violent and merciless. You can't defend nature without accepting this.

So, as I look back on that winter, the Jérôme Pensu I've become sees both my steadfast respect for animals and my evolving relationship with them. Today, I can say that I don't love animals in the singular. I love animals. By extension, I don't protect individuals, but species, and that's important.

Consequently, I believe that if mankind is to intervene in order to compensate *as little as possible* for the impact of its activities, it must also respect natural organization, and not replace it. This is my line of conduct, which, despite a thousand pressures, I intend never to deviate from.

3.
The Dream of the Pink Flamingos

A relationship is beautiful when it's reciprocal. My relationship with wild animals is one of those. Despite my beginner's mistakes, I think I've helped them since I was a teenager; in return, the animals have helped me, because, at school, I'm no longer rowing, I'm sinking. My parents are worried. Understandably so. My father comes from a working-class background. He's coming to the end of his career, having climbed many, many rungs. By dint of his dedication to his work, sometimes to the detriment of family life, he has become number three in the company that employs him. He has worked like a madman and struggled just as hard to keep his family out of trouble. This explains the pressure he puts on his children to succeed in life. Alas, earning money, starting a family, preparing for a nice retirement, being respected for what I've accumulated, far from motivating me, these prospects slide over me like water over a duck's feathers. I don't care about succeeding in life; I want to succeed in my life. And that means animals. But how do you turn a passion into a profession?

I've always loved animals. I've raised chicks, mice and peacock pigeons. I was allowed to have an aquarium that captivated me. It became my television! For hours on end, I'd watch the fish swim, eat and evolve in an environment that I wanted to be as natural as possible. When I was five, my godmother asked me what I wanted to be when I grew up.

—Farmer," I replied, linking agriculture to animals.

—Well, you have time to change your mind! she concluded, laughing.

However, years later, because of my penchant for animals and my increasingly chaotic school record, I was advised to take a vocational course in agriculture. My parents weren't enthusiastic, but they wanted me to succeed in a field I liked. The problem is that I'm not very enthusiastic either. I had a hunch that the field wouldn't suit me. In 1984, the "BTSA GNP" (Advanced Agricultural Technician's Certificate in Nature Management and Protection) didn't exist. Nevertheless, we set out to find a school that would suit me. I've set myself a red line: I don't want to get involved in an industry where the animal I've raised ends up in the butcher's shop.

I remember that day in July when we had an appointment at one of the many establishments to visit to find the one that would suit me. It's a large agricultural college in the middle of the Beauce region. To get to the principal's office, we had to cross a very, very long corridor with huge display cases crammed with stuffed birds. The man receiving us is skeptical about my project. My parents aren't farmers; I don't live on a farm; becoming a breeder would require a CAP + a BEP + considerable financial investment. After all this encouragement, he asks me:

—By the way, what do you want to raise, young man?

—This," I tell him without hesitation, pointing behind him to a stuffed pink flamingo in one of his office windows.

The guy turns around, nods and, with a slight shrug, says to my father:

—Well, at least it's clear: there's nothing I can do for your son.

When we leave the school, my parents are shocked.

—Flamingos!" says my father, unsure whether to take my exit as yet another whim or an expression of my insolence.

—Of course! What's wrong with that?

—Why not ostriches or parrots?

I smile and whisper:

—How did you guess?

*

While I'm waiting for my dream to come true, I find myself in Poitiers at a college that has just opened a CAP in animal husbandry with a "small animals" option. The aim is to learn how to raise animals such as dwarf rabbits, exotic pheasants, guinea pigs and hamsters, whose destination would be pet shops. Classes bore me. There's very little on the subject of animals. Wild animals are not on the syllabus. My little motivation disappears. I'm fired. Once again.

In 1986, I'm sixteen and out of school. For my parents, it was a catastrophe. For me, it was an opportunity to devote myself to my passion: animals. Thanks to the subsidized contracts that flourished under the Mitterrand era, I moved to Grenoble, where my grandmother lived, to take up ornithology. After that, I became a conscientious objector for two years. It was an opportunity to get my foot in the door of an association dedicated to animal protection.

The first year, I was assigned to a vulture reintroduction project. I was proud and happy. I'm responsible for some fifteen of Europe's largest birds of prey. I have everything to discover, everything to learn. Unfortunately, the party has been disrupted by the dishonesty and unpreparedness of the association that hired me. The site where the vultures are to be released is crossed by a high-voltage power line, a death trap for the birds of prey. What's more, I'm forbidden to contact the person in charge of a similar project in the neighboring département, on the pretext that "each to his own, and the vultures will be well looked after". After some vigorous clarification, I left for Brittany for my second year of objection, disgusted but determined to persevere.

Assigned to manage a nature reserve, I begin to understand what "nature protection" really means. I interact with my scientific colleagues and with the traffic section of the World Wild Fund for Nature (WWF). I deepen my knowledge of exotic birds, their breeding, capture and (enormous) mortality during transport. I become aware of the distinction between loving the animal you own and loving the animal you protect; and I have a revelation: what I want to do has nothing to do with breeding. If, one day, I have to have captive birds, it will be solely for their conservation. Above all, it became clear to me that, to protect nature effectively, it's not enough to love it: you have to know the relevant legislation. As soon as I understood this, I discovered a passion for these texts.

It's doubly bizarre. On the one hand, the texts in question are arduous, dense and jargonous to the highest degree—yes, that's right, unbearable—without anyone knowing whether they're twisted for the sake of precision or so that no one understands them, even though I have my own ideas on the subject. On the other hand, I don't like "reading books", i.e. novels. I'm incapable of it. I would later learn that I suffer from aphantasia. Simply put, my mind doesn't make or keep images. So when I hear a story about a beautiful mountain landscape, for example, not only am I unable to visualize it, but by the end of the first chapter I've forgotten what it's all about. Under these conditions, it's hard to enjoy a story...

So I only read books and technical articles about animals, but also, and above all, about legislation. I devour these texts. I think this passion stems from something my father used to say to me when faced with a math exercise: "You can't do the exercise if you don't understand the statement". Very often, in a statement, there are a whole bunch of terms that serve no purpose. You just have to ignore them to make the statement digestible. For example, in my school days, I often came across exercises that began with "either". Either a train leaving Marseille at 10.39 a.m... Or a bathtub filling up at the rate of 30 liters at an hour... And that damned "either" prevented me from understanding the exercise because, for me, this word was only used in everyday language as an alternative mark: it's "either" in the morning, "either" in the evening. If I removed the "either" from the statement, the problem seemed much less problematic!

In the same way, for me, legal texts are like a mathematical exercise that has been painstakingly made infinitely complicated—with a certain talent. Conventions, regulations and decrees are riddled with useless words. To decipher these statements, you have to spot the useless terms and ignore them. Those that remain then begin to resemble the rules of the game, which must be assimilated before the game can begin.

*

In 1991, armed with my new knowledge of the law, I decided to set up a new association to combat the trafficking of psittacines (parakeets, parrots, cockatoos, etc.). At that point in my life, I was sure that those who broke the law did so out of ignorance. I quickly realized that no, what motivates sellers is the lure of profit. They're not ignoring the law: they're circumventing it and have no intention of obeying it. That's why I've decided to fight the illegal bird trade. In concrete terms, when I spot stores that are breaking the law, I pass on the information to what is then called the Office national de la chasse et de la faune sauvage (ONCFS). For example, I tell them that a particular species should not be sold. It's up to the authorities to intervene.

All the more so since, in theory, since 1976, you can't hold a wild animal until you've obtained two authorizations: the certificat de capacité, which attests to the skills acquired by the candidate, and the autorisation d'ouverture, which aims to ensure that the facilities are suitable for holding wild animals in captivity. The restriction therefore only concerns the holding of animals, insofar as, in theory, the seller holds these authorizations. In fact, at the time, thousands of French people were keeping exotic animals (birds, turtles, snakes, insects, etc.) sold legally in pet shops, but often kept illegally. This customary infringement of the law is compounded by weak regulations. There is no system for identifying animals, whether exotic or domestic. As a result, to claim to be in compliance with the law, all you have to do is present an invoice with an animal that corresponds to the heading, even if the transaction does not involve the individual in question. It's like buying a 4L without a license plate and having an invoice that could apply to any other 4L! A single invoice can justify the purchase of several animals, opening the door to trafficking.

In this context, to curb trafficking, I can't report an offence relating to a particular individual that I suspect to have been imported illegally; I can only report a more general offence concerning the species being marketed by a particular store. Is this species authorized for sale? Is it not? Then you have to act. Simple, isn't it? Actually, it's very simple, but it's as simple as it is complicated. So it's very complicated.

Take Amazons, for example. These magnificent parrots are massively captured in the forests of South America to be sold in European pet shops. This is legal for certain species of Amazon... but the word "Amazon" covers dozens of species and as many subspecies. In simple terms, if the animal has a blue forehead, it may be sold; if its forehead is orange, but with a little blue, it may not. In practice, differentiation comes down to a few feathers on the head or tail. To com plicate the whole, plumage can be affected by stress, by the degree of acceptance of ownership and by physical condition.

In other words, to distinguish at a glance between marketable and non-marketable species, you need a habit that not everyone has. I've developed it. For a franc or two, I duplicate the litigious inserts and send them to the Office, which intervenes with great success. So much so that one day, one of the agents I worked with the most said to me:

—Your information is very kind, but it forces us to seize lots of animals.

—What's that got to do with us?

—Well, what do we do with them?

And then, blank. I don't know what to say. Sometimes it's better.

*

The agent's question changed the course of my career and my life. Until now, I had been:

- sensitive to the environment;
- passionate about birds;
- angry at trafficking that didn't even try to hide.

Nevertheless, this emotion, this desire to be useful, this conviction that I had to get involved remained vague. They weren't rooted enough in the reality of the problems, and weren't focused enough to claim long-term effectiveness. Without realizing it, the agent, who shared with me his dismay at the impossibility of actually catching the birds, decided the rest of my career. Indeed, his question pointed to one of the driving forces behind wildlife trafficking in France. Even when seizure is obvious, what con siderable solution can be proposed for seized animals? After checking, the answer can be summed up in six letters and a single word: none. This is why, more often than not, agents carry out a fictitious seizure. In other words, they don't seize what they should have seized. The animals remain with the offender, without any obligation being imposed.

Not because the officials condone this traffic, just because, even with the best will in the world, they don't know who to entrust with the animals they almost saved.

In 1991, the rare birds that were seized were done so at points of entry. When, at Roissy or in the ports of Le Havre or Nantes, customs officers intercept animals of illegal origin or species, they seize and improvise. Sometimes, they leave the animals in the care of relatives lacking authorization or capacity. In this way, they add offences to the initial offence, without resolving the problem of the destination of the animals seized. Sometimes, they hand over part of their findings to the zoos, which rub their hands and quickly realize how much they stand to gain from the process. No ill will on the part of the officials, no culpable complicity: they simply have no other option.

This impasse is one of the public impediments that most fuels wildlife trafficking in France, against the backdrop of the unfounded and dangerous myth that seized wild animals have no other destination than captivity. Assuming that release is impossible justifies keeping animals prisoner. And there's no mistaking the fact that those who insistently spread this rumor to the ears of a few sympathetic civil servants are the ones who profit most from it. As someone who began my professional life with a project to reintroduce vultures after breeding in captivity, I know how false this assertion is. From now on, I'll never stop fighting to prove it.

4.
Bébel's Parrot

The wildlife trade thrives on lies, a chronic lack of resources, a high level of ill will and the insane energy expended to discourage people who, like me, want to change a flawed system that suits everyone.

No matter how strong an animal lover's convictions, it's almost easy—as much through the aura of authority as through the number, violence or repetition of pressures—to convince them that they're wrong, that their attitude is counter-productive and that, in believing they're defending a cause that's undoubtedly noble in principle, they're actually harming it.

For me, this usually has the opposite effect. Insidious or frontal attacks stimulate me more than they discourage me. When I'm sure I'm in the right, it's hard to bend me. I'm not overpowered, I don't have a network that makes me untouchable, but I do have one indestructible flaw: I'm stubborn. Very stubborn. So that, after my exchange with the ONCFS agent, I imagined, I cogitated, I studied, I consulted; and, for several years, I worked out in minute detail how to set up a reception center for seized parrots.

The reason is both personal and obvious. Personal, because I fell in love with the bird when I saw Jean-Paul Belmondo give Claudia Cardinale a blue and yellow macaw in the film *Cartouche*. However, the fundamental reason lies elsewhere. At the time, the parrot was the most widely sold bird, and therefore the most trafficked. Animal trafficking had to be tackled, starting with what was most profitable for traffickers. A little *hitch*: this time, it's not easy at all. It's infinitely necessary, but infinitely complicated. But it's worth the effort. The only way to curb trafficking, if not put a stop to it, is to hit where it hurts: the wallet. With this in mind, illegal birds need to be genuinely seized and removed from the commercial circuit.

So I start writing my project like others would write a novel. I detail:

- the values behind it;
- the objectives it pursues;
- and the partners who want to get involved.

The amount of work is considerable and motivating. I feel I've found my calling.

*

In 1991, I finished my second year of conscientious objection. I answered an advertisement for a one-week replacement position as a marine educator on the island of Berder. I'd be there for seven years. But I didn't lose sight of my project. I soon realized that the island would be the ideal place to set up a refuge. It's the largest private island in the Gulf of Morbihan. It belongs to a religious congregation, the Little Sisters of St. Francis of Assisi, who have leased it to the LVT (Loisirs, Vacances, Tourisme) association, which is involved in social tourism. On this extraordinary site, LVT welcomes families, even very modest ones, as well as sea classes and seminars.

I arrive on the island at the beginning of what seems to be the episode of a bad soap opera. The Yves Rocher real estate company is in the process of buying the island through an estate agent. They want to get rid of the occupants and turn the island into a luxury thalassotherapy center. Pressure mounted on the prefecture, the safety commission, the town hall... Dozens of health inspections were carried out against the association that had hired me, as well as more than six tax inspections, cancellations of opening permits and defamation campaigns in the regional daily press—Yves Rocher was in fact convicted of this offence.

It was in Berder that I learned the meaning of the words pressure, administrative power, treachery and political scheming, but also combat, justice, tenacity, honor and resistance, and we'll need it.

In our battle against the behemoth, we are often alone, even if we sometimes have the support of a few motley media outlets like TF1 and *Charlie Hebdo*. By sticking to our guns, we've shattered the adage of the pot of earth versus the pot of iron, rewriting an episode in the David versus Goliath saga in the process. After seven years of fighting—and that's not too strong a word—LVT was allowed to stay on the island until the end of its lease.

Once the victory was won, the association's management understood my desire to go elsewhere, recognized my commitment to these years of struggle and validated the end of my contract. As a result, I'm now eligible for unemployment benefit, just long enough to put into practice my project for a refuge for parrots which, despite all these years, I've never lost sight of. First step: find a site with three requirements. The site must be:

- isolated, as these birds are very noisy;
- discreet, because they're expensive;
- and, if possible, beautiful, because I have to live there.

I searched for a long time before coming across a two-hectare property. "Beautiful" is not the word that, as it stands, best characterizes the place. It's not a pretty manor house surrounded by a formal garden, with a high gate to protect it from prying eyes. No, as luck would have it for my project, hidden away in the remote countryside of the Vannes region is a fibrocement pigsty, abandoned since it went bankrupt some twenty years ago. For some strange reason (unless it's to avoid conflicts with potential neighbors?), it's located in the middle of the forest.

When I arrive at the site, it's impossible to imagine the presence of the building, so much has nature reclaimed its rights. Ninety linear meters of undergrowth block my view, preventing me from entering the wasteland. I don't mind, though, as I intend to dismantle the building. I'm rather taken with the place, but I've learned to be wary of "crushes" that make you lose your lucidity. The stakes are too high for me to rely on a first impression. Many criteria have to be met. As a result, I return to see the place again and again. I check whether there are any roads that could disturb the birds, and whether people are passing nearby.

Visits and counter-visits are conclusive, I take out a loan and buy. It wasn't easy, though, because when I apply for planning permission for a parrot farm, it's a nuisance. Assuming that there are vi son, bison or ostrich farms in France, I don't see why there shouldn't be a parrot farm. Nevertheless, I need a good knowledge of agricultural and town-planning regulations to win my case. I'm aware of the risk I'm taking—professional, financial and human—but I'm certain I'm fighting a battle that's worth the effort and the cost. I believe I'm taking on a gigantic challenge. I'm about to discover that the epithet is far too weak. Luckily, perhaps, I have only the faintest idea of the trials and tribulations that await me. It's my fault, too: I want to be beyond reproach. I've seen too many rascals cross the yellow line because "it's more convenient" (which is true) or "it's not so bad" (which is debatable). I don't want to play that game. I have the firm and definitive intention not to open my establishment without having obtained all the necessary authorizations and clearances.

I estimate that it will take me a year to complete the file. It doesn't seem that long to me. I'm a patient man. I didn't expect to have to be so patient for so long. It'll take me five years of proceedings, during which I'll have to endure a dozen of them:

- unjustified refusals;
- negative opinions from pseudo-experts at the Muséum national d'Histoire naturelle, in this case sold out to the captivity industry;
- and a dozen or so inspections by government departments, set in motion by Jacques W., a senior civil servant I'd had the bad idea of meeting a few years earlier...

*

My story with Jacques W. began in 1994. I'm in the middle of planning my seized parrot center. I want to con nue the regulatory context that applies and identify the people who oversee this legal arsenal. I understand that it's the Bureau de la protection de la nature et des paysages where it all happens. I manage to get an appointment in the capital with the famous Jacques W., a senior civil servant who will go on to become Inspector General of Veterinary Public Health—no mean feat. I'm delighted to be able to fine-tune the concepts I've already acquired and the practices I plan to implement with this leading figure. I expect him to encourage me, if need be, and to point out possible weaknesses in the dossier I present to him. Without a moment's hesitation, the man dismisses my project with a wave of his hand: "A center for seized parrots? What's it for? Trafficking *is peanuts*, and the few seized birds don't pose a problem. Your center is pointless.

He was quick to add another layer in writing. On December 22, 1994[1], he wished me a Merry Christmas in his own way by sending me this letter from the "Sous-direction de la chasse, de la faune et de la flore sauvages":

1. https://www.sauvage.bzh/sommaire/?file=Chapitre04/CourrierJW.pdf

Sir,

You have sent us a detailed dossier on your project to set up a psittacine species conservatory, a facility designed to house psittacines seized because of their irregular status with regard to animal species protection provisions. [...]

However, the need for such a financial investment has not been demonstrated. In fact, the placement of seized animals pending a court decision does not appear [...] insurmountable [...]. It does not therefore appear essential to set up a facility such as the one envisaged by your association at the present time.

I could be devastated. Except that I didn't set up this project on the basis of an intimate conviction, but on the basis of field observations. I'm beginning to understand the dichotomy between the position of administrative agents and reality. That's what prompted me to take action. In the thick of it all, it's obvious that if, well if, three times if, we need centers to receive seized birds and thus curb trafficking. Admittedly, *in hindsight,* I'm surprised at my naivety. Thanks to her, or because of her, I was far, far from imagining that I was embarking on a long-term battle.

*

Some time after meeting Jacques W., I land another important appointment—this time in Geneva. The plane trip from Vannes costs me a fortune, but it takes what it takes. I've come to meet Jean-Patrick L. He works for CITES (Convention on International Trade in Endangered Species of Wild Fauna and Flora, otherwise known as the Washington Convention), which governs the conditions under which endangered wild animals and plants can be traded. This time, the contact is good, the advice precise. The man is full of confidence. He smokes cigarette after cigarette, using the last one to light the next. I tell him about my project, and in particular my concern about the fate of birds belonging to unprotected species that have been seized because they were sold in excess of authorized quotas.

—What the hell, sell them! he retorts point-blank.

Then he explains:

—On the one hand, you collect animals of species banned from sale. You raise them in the hope of one day sending their offspring back into the wild, right? Right, and on the other, you have the animals authorized for sale that were given to you because the quotas had been exceeded. The sale of these species is legal. It will enable you to finance conservation, food and staff. So sell them, because money is everything!

I like the frankness of my interlocutor. I always like the frankness of my interlocutors when they know what they're talking about, even if I'm challenged by what he's saying. The guy's got common sense and his opinions are far from stupid. And yet, when he left CITES to become a ministerial advisor to Dominique Voynet and then Yves Cochet, he never stopped putting obstacles in my way. Still under the good impression he'd made on me, it took me a while to admit that this guy was playing a double game... perhaps for a question of *ego*? Indeed, in 2000, I signed a rigorous study on the trafficking of psittacines via the Internet. The document was taken up by numerous international associations, and even presented to the British Parliament as part of a symposium on the subject. It also led me to take part in a conference at the Muséum national d'Histoire naturelle to present the center's project and talk about parrot trafficking. Up until then, the field had been the preserve of Jean-Patrick L. Did he see me as a competitor? Only one thing is certain: his hostility towards me, after his departure from CITES, and his closeness to Jacques W. will get me into serious trouble with my supervisory authorities.

This is one of the reasons why the fight against trafficking is so tough: it's not just the traffickers who are taking a beating. Those who, like you, are supposed to be fighting against their misdeeds can go out of their way to stab you in the back. In fact, I soon become the target of civil servants, between defamation and administrative blocking. To contain the danger embodied by an environmentalist, there are discreet and effective strategies such as:

- suggesting that local authorities keep a watchful eye on the unruly individual, now suspected of animal trafficking;
- repeatedly carry out checks on him;
- postponing his application for the compulsory certificate of competence until he formally threatens to sue those responsible for these postponements in the administrative court;
- or to delay the issue of the indispensable opening authorization.

These strategies are obviously cumulative. It's all about massively bombarding this nuisance, pounding his defences, in short, showing him that if he wants war, he'll get it. To make his life miserable is to hope that these rebuffs will lead him to abandon his project or, better still, to commit the fatal error of anticipating and finding himself outlawed.

Much to the chagrin of my opponents, I held firm. In 2002, my tenacity paid off. I could open. At last. When I obtained the required authorizations, Jacques W. made sure that no animal ever came to my center. What better demonstration that there is no trafficking than to leave a center in charge of collecting seized birds without any animals... and to favour hastily improvised structures to place them there—without control, without supervision and without evaluation of their future. I realize that there are two main reasons for the scale of the traffic: the power of the traffickers and the complicity of certain civil servants for reasons that don't seem to be primarily pecuniary. But why spend so much energy trying to harm me? And, if it's not just a question of money, who benefits from the huge windfall of seized birds? The rest of my story will answer that question.

5. Erika, Prestige and All the Others

It's been three years since I moved into the reception center for seized parrots. On the one hand, the stubborn hostility of the ministerial administration; on the other, my certainty that this structure is useful. While waiting for the birds, I continued to work with certain environmental police officers, regularly providing them with technical documents or taking part in expert appraisals to support their inspections. However, in the winter of 1999, an ecological disaster of international proportions turned my life upside down, while adding another string to my bow.

While the eyes of the world were focused on the world's computer systems, waiting for a possible Y2K bug, I spent Christmas with my family in the Paris region, in the land of my first ornithological experiences.

It's been 10 days since the oil tanker *Erika* sank off the coast of Brittany. A large oil slick, albeit under surveillance, is heading for the Ile d'Yeu, where all anti-pollution resources have been dispatched. The last oil disaster to hit the Breton coast was in 1978, when the supertanker *Amoco Cadiz* spilled 227,000 tonnes of crude oil. By comparison, twenty-one years on, the entire community is watching the drift of this slick with an almost detached eye, clearly failing to grasp the scale of what's in store. To be frank, a slick of 12,000 tonnes compared with the 227,000 tonnes of the *Amoco* is hardly the catastrophe of the century. In fact, that's what Dominique Voynet, then Minister for the Environment, said, a statement no doubt whispered in her ear by "specialists on the subject"[2].

But on the night of December 24 to 25, a storm of rare violence changed all that... Bretagne vivant, the association I worked for during my second year of conscientious objection, organized a network of bird collectors along the Morbihan coast. It was thinking of helping to transport a few oiled birds to the LPO center on Île-Grande, the organization's flagship site, created in the wake of the 1978 oil spill. Its members could not have imagined that they would in fact be behind the opening of the largest temporary care facility ever created during such a spill. The premises of the nature reserve, which they manage, were soon overwhelmed by the dozens of birds arriving in cardboard boxes. On the evening of the 24th, they enlisted the help of the commune of Theix, near Vannes, so that a room could be temporarily made available to them. So it was that the Salle de l'Hermine was opened to them to store the di zaines of birds. By the evening of December 25, the number had risen to 6,000.

It would take too long to recount the whole episode, which will perhaps give rise to a second book, who knows. In any case, the experience is tinged with disillusionment, self-interest and scandal, showing once again the true face of humanity.

It was during the *Erika* disaster that Breton businessman François Pinault mobilized his personal funds to compensate the

2. https://www.ouest-france.fr/pays-de-la-loire/nantes-44000/dossier-erika-il-y-a-vingt-ans-la-maree-noire-et-la-boulette-de-la-ministre-voynet-24a2e-c9e-233f-11ea-924c-4926b5fba0de

associations for their mobilization. He understood the need to anticipate a possible recurrence and decided to finance an appropriate intervention tool. I'll be in charge of supervising the design. Drawing on my new experience, and with the help of his personal architect, I design and develop a mobile care unit for oiled birds (UMSOM). Unfortunately, it didn't take us long to demonstrate its usefulness.

In December 2003, just three years after the *Erika*, the *Prestige* sank off the coast of Galicia. At the same time, several ships ran aground in the North Sea, causing the greatest pollution in terms of coastlines affected. With the exception of the Mediterranean, all French coasts are affected by pollution. A unique alliance was forged with the Landes hunters, the only organization in Aquitaine to agree to host the UMSOM during this oil spill. The *Erika* left its mark, and not just on the rocks, and certain associations such as the LPO, noting their loss of supremacy, take a dim view of the arrival of this new tool on their turf.

The hunters, the only association to have accepted the arrival of the UMSOM, will install us in a pheasant breeding park, empty during the winter period and usually home to several thousand individuals destined for hunting. Yes, I know, it's a funny situation, but in times of crisis you have to know how to make aty piques alliances for the benefit of the cause.

For me, this is an opportunity to prove that preparation, anticipation and training can reduce intervention costs and optimize rehabilitation results. Well-prepared, we'll be releasing over 30% of the birds entrusted to us—double the usual rate in such situations.

*

During the *Prestige* oil spill, I realized that there were no wildlife care facilities in Aquitaine, with the exception of a tiny, isolated establishment deep in the Pyrenees. On the land adjacent to the breeding park where the UMSOM is located, an old building sits almost abandoned. Every day, on my way to the site of the breeding park, I pass by this surprisingly shaped building (supposed to represent a boat), built to house a hunting museum. On enquiry with the president of the hunting federation, the museum is only open to a limited number of curious visitors. Its location, some fifteen kilometers west of Mont-de-Marsan, far from any tourist area, does not encourage visitors.

—So we're going to cut it down," he tells me. We've released a €300,000 loan to bring the building down.

I jump and retort:

—Do you know what you can do with that kind of money?

A questioning glance. I insist:

—Remember! Three years ago, you were complaining that there was no health center in the area. You keep reminding me that, statutorily, the Landes hunters' federation is an association for the protection of nature.

—This is true.

—Then prove it.

—How can I do that?

—Give me your €300,000, and I'll build you a care center.

So, admittedly, for the general public, the hunter evokes more the killing of Bambi's mother than the care given to an animal in distress. In reality, the animals that may be taken in at a center are not huntable. In the vast majority of cases, they belong to protected species. Treating them therefore does not harm hunters. Above all, it has to be said that Landes hunters seem to me to be rather atypical. They were massively involved during the oil spills. They manage several reserves where hunting is banned and where many emblematic species live, including cranes, storks and ospreys passing through. Even as someone who doesn't usually spawn with their kind, I can see that their relationship with nature doesn't stop at hunting itself.

A final argument for considering working with them: the departmental president of their federation is renowned for expending considerable energy to bring to fruition the projects that are close to his heart. As it happens, the idea of a care center has been on his mind for several years. A first attempt was made by the federation just after its investment in the *Erika*. However, given the complexity of the issue, especially for non-specialists, and the hostility of certain supervisory authorities, associations and other nefarious parties, the matter came to nothing. My proposal didn't fall on deaf ears, even if the man was wary.

—You know, he murmurs, we tried it, and it didn't work. No offense, but there's no reason it should work out any better with you. To be honest, I don't even think it's possible...

Did he do it on purpose? He's just uttered the magic phrase for motivating a Breton. I smile and resume:

—Give me the €300,000, I tell you, and I'll build a health center on that land!

We hit the ground running, and events accelerate. Within a fortnight, the president had the federation's board of directors validate the principle of transforming the museum into a care center. I drew up a feasibility study, consulted the federation of care centers of which I had become a director six months earlier, and set about listing the needs of the future structure. The results were crystal clear. The Landes département is very large. Pouydesseaux, where the hunting museum used to stand, is located in the middle of Aquitaine, which makes it possible to collect animals from just about anywhere. Robins, cranes (of which there are many in the département), grey herons, birds of prey, foxes, martens, otters and marine animals could all be cared for here. Clearly, a care center is fully justified!

I translate this certainty into technical terms. My work convinced my clients. They extended my mission to the realization of the facility. In April 2004, I was hired by the Landes hunting federation. While returning to Brittany at weekends to look after my animals, I threw myself into this new challenge.

*

All around me, teeth are gnashing. They call me a sell-out, even a collaborator. I'm in bed with the species most hated by countless conservationists. And yet, my involvement with hunters advances the cause of ecology. First of all, I'm turning a hunting museum into a health center—that's quite a story, isn't it? Secondly, I think it's very healthy to set up beautiful projects with people whose convictions I don't share, and vice versa. For me, exchanging and even building with those who are not exactly of my persuasion is not selling one's soul, it's:

- coming together around good ideas;
- forging constructive alliances;
- taking advantage of a combination of energies and personalities to defend wildlife together, even if we don't agree on everything!

I'm not interested in pleading to an acquired public; I'd rather convince the skeptics. But I'm no longer naive. I'm beginning to have a bit of experience. I know the fluctuations of ethics in *homo sapiens*, hunter or not. So, right from the outset, I set out a triple and definitive rule, which allows us to join the French Union of Wildlife Rescue Centers, which is usually hostile to hunters. This triple rule stipulates that:

- the head of the center, who must hold a certificate of competence, must not be a hunter;
- the center will welcome all animals, including what hunters call "harmful" animals such as foxes, badgers and martens;

- all animals, including game (with the exception of roe deer and wild boar, whose care is too problematic), will be cared for exclusively with a view to release.

The Alca Torda adventure—from the Latin name of Europe's rarest penguin, and the most difficult to save during an oil spill—is beginning to take shape. I'm both project manager and project owner by delegation. Months of hard work are finally paying off. When all was accomplished and Henri Emmanuelli, President of the Conseil Général, inaugurated the center, my mission was over. I could return to Brittany while keeping one foot on place through my involvement with the "Captive Wildlife" commission of the Landes prefecture. The local press is amused by this unexpected association between hunters and ecologists, with the headline: "Alca Torda, the ecological PACS"!

Even though I'm aware that I'm leaving behind a wonderful tool, I have no regrets about handing over management of the center to another holder of the certificate of competence. He will run it from 2006 to 2010. Helping to protect wildlife by treating animals is not my life's ambition. I'm more driven by my desire to fight trafficking from my property in Brittany. Year in, year out, I remain convinced that there is a way of developing a reception center for seized parrots, thanks in particular to the contacts I've made with the Landes region and the French Union of Wildlife Rescue Centers. By nature, I'm resolute and far more optimistic than frowning. The harder the fall, the better!

6.
The Law of *Lobbies*

The parenthesis "managing a large-scale disaster" is closed. The one devoted to the invention of two care centers (mobile and fixed) is also closed, and I return to my primary mission: trafficking and traffickers.

As I've written, the traffickers would be nothing without the high officials, notably from the Ministry of Ecology. For the sake of clarity, I'd just like to make one point about this aspect of the drama. To this end, we need to be aware that, when a politician has decided to pass a law, there is no guarantee that, *in the end,* the text will go in the direction he or she wishes; and the craziest thing is that he or she is fully aware of this.

An example: the November 30, 2021 law "aimed at combating animal mistreatment and strengthening the bond between animals and humans" was intended, among other things, to put an end to wild animal shows in circuses and dolphinariums. First subtlety: zoo shows are spared, for those who might have doubts about the power of the captivity industry. I attended the creation of this text. On numerous occasions, I was interviewed by members of parliament and by officials from the Ministry of Ecology. The latter are the real conductors of the orchestra. From the outset, they set the framework for what will be acceptable to the Ministry and what won't, which poses a problem in terms of democracy. In practice, every time I made a comment, I was told: "No, that office will never let that through".

So, contrary to appearances, it's neither the minister nor the parliamentarians who choose the contours of the law. But above all, the final text will carry a clause stating that "an order of the Minister will specify the conditions of application of the law", and this is where the *lobbies* take over.

Four years after the text came out, a multitude of derogation regimes will enable dolphinariums to continue their activities, since arrangements will be put in place to do so. The famous pretexts of scientific study, pedagogy and protection will be included as a pledge of seriousness to ensure the continuity of animal exploitation. The members of parliament are happy to have brought forward a supposedly innovative and courageous text, and the *lobbies* are happy to have obtained a derogatory arrangement allowing nothing to change, even if the media tell us otherwise.

It's on this principle that senior civil servants have been deconstructing the law for the past 30 years, or issuing decrees which, by means of derogations, allow traffic to continue despite a regulatory framework which everyone believes to be increasingly strict.

In this sense, parliamentarians and politicians in general have long since lost the democratic power conferred on them by their election and delegated to them by the electorate, leaving an army of civil servants to take over, all the more so since the highest-ranking officials can sign the decrees they draft, many of them having been delegated to do so by the Minister. When you see this kind of drift, you know that France has become nothing more than an administrative monarchy, where the power of voters and their representatives is reduced, with rare exceptions, to mere representativeness. This is the basis on which the whole context of wildlife trafficking in France rests.

*

In this misuse of power, 2006 marks a turning point.

Until then, the pressure exerted by captivity *lobbies* on their supervisory ministry had never been so strong. This *lobby* brings together two branches of the *business*: zoological parks and associations of exotic bird breeders. One day, a small decree was presented to the Conseil national de protection de la nature (CNPN). Those who know how to read it (and they are rare) can measure the degree of entrism and the power of influence of the *lobbyists*. The text lists the species of French wildlife that may not be kept, which is laudable. However, it specifies that this ban applies "unless the animals were born in captivity". This "except" changes everything, for one obvious reason: there are no effective measures in place to distinguish between birds born in captivity, which are legal to keep, and those taken from the wild, which are illegal to keep. In other words, by introducing this derogatory regime, the officials behind the text are opening the door wide to trafficking. They allow the mass regularization of all birds illegally held by breeders and zoo owners. This marks the start of an unprecedented money-laundering operation.

The technique is simple. All over Europe, individuals, breeders and zoos were clandestinely holding birds captured on the continent, which in France is forbidden. In the 1980s and 1990s, this appropriation of wildlife was all the rage. It affected ducks, swans, geese and more. The Belgians, Dutch and English specialized in keeping robins, tits, woodpeckers and birds of prey. I insist: in France, since 1976 and until 2006, this has been completely impossible, forbidden and illegal. To compensate for this gap between legal impossibility and lucrative reality in the field, the 2006 decree rolls out the red carpet for administrative laundering.

Indeed, since 2004, to be legal, a bird must be fitted with a specific marking (closed ring or chip called a "transponder") combined with an entry in a special register and a certificate of transfer. The aim is to ensure that captive birds come from legal sources. Bands are essential, because to prove that a bird was born in captivity, they must be put on within fifteen to twenty days of the chick's birth. It will grow up with its band so that, normally (the adverb is important), removing it is impossible. Consequently, a banded bird is considered to have been born in captivity. Unfortunately, this marking system can be falsified at will. Only a thorough check and an expert eye can detect tampering. As I write these lines, the Parc des Pyrénées is being tried by the Cahors magistrates' court for trafficking in wild birds involving, among others, breeders, zoos and an expert from the Muséum d'Histoire Naturelle. The whole fraud used to launder illegally caught species in the wild is directly induced by the safe-conducts laid down in the regulations.[3]

When it was forbidden to keep European birds, controls were simple. By authorizing their possession "if they were born in captivity", the authors of the decree have opened a Pandora's box that will no doubt be impossible to close. All the more so as a few subtleties in the decree open up other possibilities for blanking. For example, it is enough to make a false declaration of birth in captivity, then mark a bird by disguising its origin, and that's it. Henceforth, traffickers are free to capture chicks in the wild, band them with a foreign ring, associate them with a German or Dutch certificate of transfer, and nobody will be able to prove the fraud, since the French authorities rarely have the means to verify an extraterritorial origin. Incredible as it may seem, a simple attestation of an animal's provenance (in this case, a certificate of transfer), whether Belgian,

3. https://www.assemblee-nationale.fr/13/rap-info/i3142.asp

German, Dutch, English or Spanish, legalizes the possession of a bird on French territory.

This technique makes it possible to launder stocks of birds which, in turn, will produce generations of illegal animals, but legally marketed. All the more so as the 2004 text authorizes the marking of European birds with a microchip rather than a ring. While a ring requires a young bird to be in hand, i.e. supposedly born in captivity, a chip can be applied to an adult bird that may have been caught in the wild. And that's not all! Traffickers can also buy rings in a bird-producing country, or recover rings from dead legal birds and put them back on illegal animals with complete impunity. They also have free rein to enlarge or shorten bands, and declare more young than were actually born, or even reproductions from two males or two females. Who's going to check?

The most frightening aspect of this story is that, before publication, these texts were presented to the CNPN, where an army of associations and experts sit. Not a single one of them raised a finger to protest against the euphemistic risk of drift presented by the text. Not one... except me. To no avail.

*

By making it so easy for the crooks, the French government has given the wild bird trade a free hand. The captive bird *businessmen* are rushing to cash in on this timely gift. In 2005, a powerful avian flu epidemic led Europe to halt imports of non-EU birds. Zoos and breeders found themselves *de facto* short of animals. They'll have to find another source of food, because *business* depends on it. As a precautionary measure, since traffickers can never have enough, the zoos put pressure on the State to ask its agents for flexibility in the event of an "administrative error". The offenders' argument is clear: fighting trafficking is an absolute necessity, a *top* priority, an essential mission, BUT, we're not traffickers. We pay taxes, we contribute to conservation, we create jobs in the regions... So, yes, sometimes there may be a strange or missing piece of paper to justify the provenance of a particular animal; but that's just an administrative problem...

Of course!

At the same time, it's not untrue. Trafficking is, in essence, an administrative problem. This applies to all trafficking. For example, a weapon may be legal, but if you don't have a permit to carry or possess it, you're outlawed. Likewise, a car may be legal in itself; but if your license plate, registration document and chassis number don't match the vehicle, it's an administrative problem whose result is that you're outlawed. We could go on and on—what's the point? It's obvious that what some people call an "administrative problem" is nothing more than the result of trafficking; and the circus, zoo and breeding world is no stranger to these more or less skilful attempts to get around legality, some of which might seem comical if they weren't appalling evidence of an illegal transaction, which finds sympathetic support among high-ranking civil servants.

In 2004, *lobbying* led to the publication of two scurrilous decrees. Everyone got their money's worth. Breeders' *lobbies* succeeded in making the keeping of animals by private individuals much more flexible; zoo *lobbies* obtained the same measure for breeding establishments, public displays, animal rental establishments, and so on. After these two blows, the keeping of wild animals is simplified and the possibilities of control are complicated. Because standards are cancelled and illegal transfers can be legalized, a large part of the traffic is transformed into a legal trade. This is nothing less than money laundering.

Should we despair? Not entirely. Apart from death, promised to each and every one of us, there is no fatality. Trafficking in wild animals in France is a tragedy that can be fought. On one condition:

don't give full power to the *lobbies*. As long as the *lobbies* dictate not *their* law, but *the law*, modifying it as they see fit, and obtaining accommodations when they are caught red-handed, the fight is impossible. As long as they have the support of the most senior environmental officials, hindering their destructive capacities is a fantasy. In the past, when animal trafficking was a major global market, France found it difficult to curb it on its own territory. Gradually, thanks to checks carried out at major Paris airports, a few high-profile cases and a few in-depth investigations, law enforcement agencies have managed to put the criminals' omnipotence into perspective. All it took was a few decrees to get traffic back into the *open-bar* spirit. So much the worse if this contradicts all the international conventions on the subject signed by France. *Business first*!

7. Whitewashing Pink Flamingos

It was against this backdrop that the veterinary services of the Landes département, who had come to know me well, approached me to inspect a very large breeding facility that was, at the time, totally clandestine. The establishment wished to comply with the new regulatory framework defined by the decree of August 10, 2004 "setting the conditions for authorization to keep animals of certain non-domestic species", before the deadline for regularization set out in the text expired. For the establishment to be authorized, the first requirement was that it had submitted an application in due form, and that a holder of the certificate of competence was present. At the time, this was not the case... even though it had been a requirement since July 10, 1976.

I have therefore been commissioned to carry out an appraisal to answer two questions: firstly, are the animals kept in conditions that respect their biological requirements? Secondly, are they legally sourced? When I received the file, I fell off my chair for the first time: it concerned hundreds of animals. Hundreds, that's a lot! I start by asking about the head of the organization. I fall back in my chair when I discover that he's supposed to be a big shot in the field. He's no more and no less than the head of the very powerful Aviornis, a European association of bird breeders, including pheasants, ducks, geese, pans and cranes, whose slogan is "*Breeding to conserve biodiversity*". This is hardly original, since "conserving biodiversity" is the preferred slogan of breeders and zoo owners wishing to justify their ownership of animals. Aware of being illegal, the man saw the 2004 decree as an opportunity to make acceptable what was only tolerated.

Intrigued, I studied in detail the elements in my possession before submitting my report to the Landes departmental directorate for the protection of populations (DDPP). In this report, I suggest requesting the assistance of the CITES/CAPTURE brigade of the environmental police, as the case is colossal and cannot be handled without this help. My suggestion was acted upon. The brigade is dispatched to the site so that we can carry out the inspection. We are greeted by the owner's wife, a particularly vindictive lady who takes offence at our presence and becomes threatening. She repeats over and over that her husband sits on the national commission that issues certificates of competence to the Ministry, and that he works regularly with Jacques W., implying that he is untouchable. She refuses to answer our questions, targets her attacks on me personally, and casts doubt (no mean feat!) on my skills. In short, she uses every means at her disposal to prevent the audit. To no avail.

After this eventful visit, we meet up again at the environmental police office in Mont-de-Marsan, to study, bird by bird, the origins, protection status, ringing, etc. of the birds. We discovered 27 Lesser Flamingos, of the species name *Phoeniconaias minor*, imported from Kenya a few years earlier. All carry a CITES certificate stating the year they were captured and imported. However, French law unambiguously states that it is forbidden to keep flamingos if they are of wild origin. We have therefore decided to enforce the law and take these illegal animals away from their owners.

A few weeks later, the 27 pink flamingos were rounded up. To carry out the operation, we can count on the Amnéville zoo, which has agreed to take in the birds. We chose this zoo for one main reason: it's located a thousand kilometers from the illegal breeding site, and therefore less likely to be influenced by the offender. We're also counting on the CITES brigade, whose mere name causes cold sweats in almost all French zoos. But their chief tells us:

—I'm glad I don't have to sign the seizure report, because this is going to be a big deal!

—Don't worry about that," replies the woman in charge of the operation. I'll take care of it.

—Do you know who you're messing with?

—Yes, but the rule must be the same for everyone. Well, *should* be the same...

Except that things aren't exactly going according to plan. When we arrive, the owner's wife, just as furious as when we first met her, calls the ministry to object to the seizure of the flamingos, her husband's birthday present. She passes the phone to the DDPP representative. On the other end of the line, one of Jacques W.'s henchmen explains that she wants to apply a dusty text that should have been amended. To which the agent retorts:

—Look, this "dusty text" was precisely dusted off eight months ago. If you don't know how to clean, learn. I apply the texts in force. What's more, we're faced with an establishment that hasn't been up to scratch since it was set up, and which is keeping animals in farcical conditions that shouldn't be kept at all, all in the absence of an accredited keeper. Quite a lot, don't you think?

—Of course, you're right in principle. However, you also know that there is the law and the spirit of the law. They're two separate things, and I'm sure there's a way to...

—All right, that's enough. Remind me, who's your responsible minister? The Ministry of the Environment?

—Mine's Agriculture. If you have a problem, call my minister.

She hangs up, and we proceed with the seizure. Then the legal machinery gets under way, a trial is held, and the counter is acquitted thanks to a little regulatory subtlety. In 2009, Jacques W. rescued his friend by passing a decree modifying the conditions for the protection of birds on French territory. It sets out:

- the list of protected birds;
- the list of birds that can be kept in captivity;
- and the conditions under which birds may be kept in captivity.

The pink flamingo *minor* has a particular fate in store for it: from now on, it may not be kept "if it has been captured in France". The risk is low since, apart from a few individuals escaped from captivity, there are no *lesser* flamingos in France. These birds live in Africa. As the capture of birds is now prohibited only on French territory, there are no grounds for prosecution. The incompetence of the animal protection associations sitting on the CNPN and the willingness of a senior civil servant to come to the aid of a trafficker have saved the accused, trampling underfoot the courageous work of the control services in the process.

*

The laundering of the pink flamingo trade is another in a long list of attacks on biodiversity signed by Jacques W., a civil servant transferred because he was caught for similar collusion in a dog trade a few decades earlier.

In the 1990s, an unfortunate car accident involved a vehicle in which dogs were found with their ears cut off—in other words, dogs that had been mutilated to remove their tattoos. The astonished gendarmes came to see the accident and decided to investigate. The investigation revealed that the animals were destined for an animal experimentation laboratory, i.e. a structure that must be run by a veterinarian. As Jacques W. is himself a veterinarian (he defended his thesis in Toulouse in 1978), it is conceivable that his network of peers contacted him. In any case, he urgently issued a circular stipulating that dogs destined for laboratories need not be identified.

The affair should have ended there, but for once media personalities committed to animal protection, notably Brigitte Bardot and Dany Saval, were moved by the trickery, and lodged a complaint. The civil servant was not disbarred. He was simply transferred. Formerly with the Ministry of Agriculture, he was transferred back to the offices of what was originally just the Secretary of State for the Environment. At the time, it was a pitiful hangout for placarded agents—which says a lot about the place of environmental protection in French politics... However, from the Mitterrand era onwards, the little storeroom, as dingy as it seemed, grew in importance. Little by little, the transfer that was supposed to be a punishment was transformed into a fine promotion. The man who had been found complicit in animal trafficking found himself in charge of the management of:

- nature reserves;
- circus issues;
- captive-animal issues;
- monitoring zoos, breeding farms and care centers, etc.

Absurdity or the pinnacle of French logic, it is he who writes the regulations and drafts the decrees setting the framework for the application of the laws. From 1990 to 2018, he worked hard—with remarkable success—to unravel the texts that attempted to limit wildlife trafficking. I've often wondered why he took on this task. My hypothesis is that he originally had a very good position at Agriculture. Thanks to a driver's clumsiness, he was caught red-handed. Big associa tions fell on him. He fell off his pedestal. I've come to imagine that he's spent the rest of his professional life taking revenge on animal rights activists, the animals themselves and therefore biodiversity. I don't believe for a moment that he committed so many misdeeds for money or by reluctantly bowing to pressure from the *lobbies*. Not at all. The man seemed neither venal nor weak. In my opinion, settling scores was his only motive; and resentment is sometimes an even more powerful motivator than sex or money.

8.
Associative Mayhem

So, are you outraged to see wildlife trafficking fed from within the system by zealous collaborators of the traffickers? Hang on, you haven't seen anything yet! There's something even more shocking. In the continuing decline of wildlife protection tools in France, the administration is not the only culprit. Its decisions are tacitly validated by the major environmental and animal protection associations. These colossal machines sit on the joint commission to provide input, their opinions and expertise to the Ministry of Ecology. As Albert Einstein once said, "the world will not be destroyed by those who do evil, but by those who stand by and watch."... And so, every time a deleterious decree has been published with the aim of inciting traffic, not a single behemoth in the sector—such as LPO (Ligue pour la protection des oiseaux), FNE (France Nature Environnement), SNPN (Société nationale de protection de la nature), to take but a few examples—has lifted a finger, or if they have, it's with such restraint that no one has noticed. And yet, their raison d'être is unambiguous:

- "stop the destruction of ecosystems" (FNE);
- "look after the interests of nature and biodiversity" (LPO);
- or "protect wild spaces and species based on the scientific foundations of ecology and conservation biology" (SNPN).

What's more, these are neither partridges of the year nor six-week-old rabbits. Each structure prides itself on its importance, be it:

- historical (the SNPN "is the leading nature protection association in France");
- financial (the LPO is "one of the largest French associations dedicated to the protection of nature");
- or societal (FNE brings together "6,206 associations for the protection of nature and the environment", representing "hundreds of thousands of activists, volunteers and employees").

In other words, each one is supposed to be able to bang its fist on the table, blow the whistle and stop the massacre, especially as their participation in the public consultation of experts can serve as a guarantee for the "adjustments" obtained by the *lobbies*: if the LPO has neither protested nor slammed the door, the regulatory changes must be good for nature, right?

The inefficiency of associations is due to two important factors, one financial, the other technical. Financially, the large associations that sit on the consultative commissions of the Ministry of Ecology depend directly or almost directly on the subsidies paid out by this same ministry, and are therefore reluctant to bite the hand that feeds them. In 2011, a parliamentary report criticized the Fondation Nicolas Hulot pour la Nature, the Ligue de Protection des Oiseaux and France Nature Environnement for their operating costs, which were respectively 50% for FNHN, 45% for LPO and 65% for FNE[4]. Under these conditions, it's easy to understand the associations' dependence on State money, and also to understand that everything is done to avoid offending the ministerial administration, in which Jacques W. works. The same Jacques W. who proposes to the

4. https://www.assemblee-nationale.fr/13/rap-info/i3142.asp

Minister not only the names of the associations that sit on the joint commissions, but also the amount of subsidies that will be allocated to them, the prey is in the trap.

The second reason for this inefficiency is ignorance of the problem of trafficking, which is dealt with by a few old opportunists to whom everyone owes allegiance despite the mediocrity of their actions and results in this field. Most of them are twenty years behind the times, and only know traffic through the prism of the 1980s. They have no knowledge of fraud and money-laundering techniques, no knowledge of the networks and organization of this traffic. Their only contacts are with the upper echelons of the environmental police force, which they present as an address book that will enable them to shine on committees or at conferences devoted to the subject.

This sometimes self-sufficient associative world has sold out its militancy, which has enabled it to remain in the good graces of the all-powerful ministerial administration, thus ensuring its presence at the various commissions, symposia and conferences organized by this same administration, in an enormous intellectual and collective masturbation that everyone echoes. What have the Universities of the Environment and the other Grenelle conferences changed? Nothing. But taking part in them is essential for these structures in need of recognition and, above all, resources.

In the meantime, an associative war is being waged behind the scenes, with each of them stepping on each other's toes to gain the honor of being received by the ministry responsible. The winner is quick to take up the other's cause and claim a place on consultative commissions or other representative bodies. This is a far cry from the collective interest, and makes it easier to understand the power and effectiveness of *lobbies*. When hunters or zoos defend a position, they have a representative who will plead a cause and a common objective. Ecological associations have as many positions as they have members, all carried by a braying herd in leather sandals, dirty beards and militant T-shirts.

Jacques W. has trained his successors, first and foremost Gérard B. Propelled to head the captive wildlife division of the French Ministry of Ecological Transition, this civil servant is capable of issuing orders to guide the application of legislation, or intervening in inspections, including judicial ones, without having the right or any official prerogative to do so. Since no politician has put an end to the deleterious operation of the W. system, the current successors continue to be intoxicated by the perverse strategy of their predecessors.

In practice, the W. & Cie system, by introducing an interminable series of exemptions which, combined with the lack of resources, training and freedom of initiative from which the environmental police suffer, and the inefficiency, not to say inaction, of the associative world, constitutes the cornerstone of wildlife trafficking in France.

This phenomenon is undoubtedly the reason why nature has never been so poorly defended. Never in the last 15 years has there been so much talk about biodiversity and its erosion, never has specific legislation been put in place like the law on the reconquest of biodiversity, and yet never before have so many species disappeared at the current rate, and never before has biodiversity suffered so much damage as a result of trafficking.

9.
The Impregnation Trick

To be complete in this inventory of the origins of trafficking, I must stop off at the wildlife care centers. Officially, these institutions take care of injured European wildlife and release them back into the wild. However, some of them regularly feed the trade by selling sharked animals to zoos. Once again, it took me many years to realize the extent of this phenomenon.

I begin to open my eyes thanks to what happens during the winter of 2014, at the Alca Torda center. Xaver, Dirk, Ulla, Godehart, Tini, Petra, Qumaria, Ruth... these are just some of the 28 storms that hit western Europe between October 26, 2013 and February 7, 2014. An impressive number of birds wash ashore. The number is equivalent to the catastrophe that would have been caused by an oil spill. Alca Torda welcomes over six hundred completely exhausted birds. During a storm, a bird stops feeding. On the high seas, it has to surf for days on end on waves measuring from two to five meters... until the wind blows it back to shore, exhausted.

Given the condition and number of birds, we set up an emergency plan following the protocol applied during oil spills. A fortnight before the first birds were due to arrive, we learned that from Biarritz to Brest, specialized seal care centers were saturated with young seals in the process of being weaned, also victims of the storms. I speak to a manager from the Réseau National d'Echouage (RNE) based in La Rochelle, in charge of managing marine mammals and turtles in distress. He confirms the news. I inform him that we have 500 m^2 of buildings and ten swimming pools available. Admittedly, we don't have the capacity to take care of seals, but it might be a good idea to include us in an emergency plan. I'm not in the mood for that kind of special treatment. However, the situation prompts me to propose it. All we would have to do would be to be granted an exemption under the supervision of the Biarritz aquarium, a recognized care center for these marine mammals. The aquarium would send us animals for rehabilitation, the simplest stage before they are released back into the wild.

My proposal was approved and we became an annex of the Biarritz aquarium. I applied to the Landes prefecture, and obtained the one and only derogation of my career, in the service of the general interest. In return, we submitted an application for authorization to open and requested a certificate of competence for one of our groomers, who was highly skilled with these animals. I then met the RNE for marine mammals. It manages the monitoring and reception of stranded animals, mainly seals and turtles. This network seems to operate in a completely hermetic way, with no connection to conventional care structures. It has to be said that on the Atlantic coast, the centers in question are all zoos, like the Biarritz aquarium, the La Rochelle aquarium or Océanopolis in Brest.

In total, we recovered seven seals... except that they didn't correspond at all to our agreement. We were supposed to recover animals that had been cared for and almost requalified, just needing to regain their weight in the pool. Instead, we end up with individuals that look more or less on the verge of agony. Not the kind of specimens you'd send to novices, however well-intentioned they may be. Why the *miscasting*? I'm inclined to think that some people took a dim view of the Landes hunters' federation obtaining approval to care for seals and being included in the RNE in the long term. By sending us the most difficult cases, it's possible that some malcontents hoped that the animals would die on our premises, to prove our incompetence.

If that was their hope, it was disappointed, because we had two trump cards up our sleeve. On the one hand, the presence of the future capacity holder, whose experience and know-how were perfectly suited to the situation. Secondly, the collaboration of a veterinarian specializing in wildlife who, shortly beforehand, came to offer us her help during the gap year she had decided to take. She offered to work for the center on a voluntary basis, in exchange for the loan of accommodation. Guess how we responded. With these two assets, we managed to save the seven seals that had been brought to us at the end of their rope. A 100% success rate for a first operation is no mean feat!

*

The shipment of dying animals by the Biarritz aquarium made me wary. Regularly, during this crisis, I scrutinized the strandings through the news that the institution gives via its Facebook account. One day, we learn that an old male seal that was the admiration of visitors to the Biarritz marine zoo has just died. I immediately understand what's going to happen and tell my team:

—I bet they're going to keep a new one to replace the dead one!

—Man, the more things go, the more paranoid you get," laughs a colleague.

Paranoid? Well, maybe. But lucid too: what followed proved that one didn't prevent the other.

Four days after the death was announced, a young seal washed up on a Biarritz beach. Even before the animal had been examined, the aquarium's communications department announced that the beaching had been a gift from heaven, and that the animal would be kept captive at Biarritz[5]... on the pretext that it had been tamed. The expert report obviously confirms this diagnosis, since the institution that commissioned the examination is the one that will subsequently exploit the animal. The Musée de la Mer, like many other aquariums, behaves like a trapper, capturing and commercially exploiting wild fauna under the pretext of saving it or helping others to discover it. This behavior seems to be tacitly validated by the RNE, since the Musée de la mer, and in particular its manager, is its representative for the South Atlantic region[6] and the technique of keeping seals in captivity under the pretext of taming them is known to all.

However, the French Office for Biodiversity, the Muséum National d'Histoire Naturelle and the French Ministry of Agriculture are all members of the RNE, and all of them validate the principle of illegal capture and maintenance in captivity, insofar as "the specialists" have assessed the animals concerned and determined that they cannot be released. They will therefore be presented to the public in the name of "education and conservation", which in the space of 10 years has become the safe-conduct that makes captivity tolerable in the eyes of the general public and justifies its principle.

Some of the seals presented in Biarritz will be exploited for more than twenty years without any authorization, in violation of all the texts protecting the species. In fact, the DREAL Aquitaine (Direction régionale de l'environnement de l'aménagement et du logement) only granted a derogation for the "definitive detention of protected animal species" at the end of 2015. Authorization 2015-149-011

---

5. https://www.aquariumbiarritz.com/fr/izar-nouvelle-etoile-dans-le-bassin-des-phoques/
6. https://www.observatoire-pelagis.cnrs.fr/wp-content/uploads/2020/12/RapportEchouages2011.pdf

specifies that one of the seals was captured in 1995 and the other in 2005[7]. With no derogation, they were exploited with impunity, one for ten years, the other for twenty, without any controls, probably under the pretext of "impregnation". For an animal to be impregnated, it must arrive at a care center at a very young age and be subjected to:

- the incompetence of a caretaker;
- the desire to keep the animal under the pretext of impregnation;
- and false impregnation, which doesn't exist, but which a complicit expert will declare to please an unethical operator.

---

7. https://www.pyrenees-atlantiques.gouv.fr/contenu/telechargement/13029/88368/file/Recueil+n°30+du+23+juillet+2015.pdf

10.
The Turtle Scam

The more I learn and the more experience I accumulate, the more I realize that trafficking often thrives under the cloak of public-interest operations. While wildlife care centers are fine projects in themselves, those run, affiliated with or financed by zoos that profit from animal captivity are, at the *very least,* a farce that engenders wildlife trafficking.

I'd like to illustrate this point with an example concerning the species of European fauna most frequently bleached by zoo care centers and marketed with the complicity of the French administration: sea turtles. In Europe, around thirty of them are kept in a dozen aquariums. Turtles:

- Loggerhead (*Caretta caretta*);
- green (*Chelonia mydas*);
- leatherback (*Dermochelys coriacea*);
- hawksbill *(Eretmochelys imbricata*).

All come from the wild and have been illegally captured under the pretext of helping them. Marineland d'Antibes even seems to have made a specialty of recovering sea turtles from the wild. Yet it claims to be a specialist in captive births, while none of the world's 29 marine zoos, holding 370 individuals, has achieved this feat.

Yet between 2010 and 2011, this marine zoo declared the birth of nine loggerhead turtles[8]. In addition to the hushed-up achievement of what would be a world first, these turtles, though supposed to have been born in captivity, are declared to have either undetermined parents, or an undetermined father. Between 2017 and 2020, Marineland captured nine loggerhead turtles, three of which were released in 2016[9], what has become of the others? Yet the Zim's record of Marineland's turtles, edited in 2020, shows that these animals are still in the aquarium.

The Marineland association has infiltrated even the French herpetological society through the Réseau Tortues Marines de Méditerranée Française (RTMMF), whose work it has co-financed in order to make the gullible forget that sea turtles are forbidden to be captured for commercial purposes, and to display its participation in a protection organization that undoubtedly makes it easier to obtain contacts and authorizations.[10]

Marineland is one of the marine zoos that have set up a care center, as have the Biarritz aquarium, Nausicaa in Boulogne, Océanopolis in Brest, the Seaquarium in Le Grau-du-Roi and the La Rochelle aquarium, whose center has taken in 186 sea turtles. In fact, where do the turtles featured in this aquarium come from?

In France, not a single sea turtle is presented to the public, in compliance with CITES rules, but everyone seems to be fine with that, since it's a question of conservation... Note that, in this field, the problem is not exclusively hexagonal. French players are simply following a global trafficking model. A case in point? Morgane, an orca captured on June 23, 2010 in the Wadden Sea, northwest of the Netherlands. She was taken in by the Harderwijk dolphinarium

8. https://www.sauvage.bzh/sommaire/?dir=Chapitre10/Tortues
9. https://www.sauvage.bzh/sommaire/?dir=Chapitre10/Tortuesrelachées
10. https://www.sauvage.bzh/sommaire/?dir=Chapitre10/RTMMF

as she was deemed “too weak, very emaciated and completely dehydrated”. She was to be kept until she was back in shape, but, surprise, surprise, the dolphinarium made sure she couldn’t be released. As soon as she arrived, the orca was placed in a “ridiculously small” tank, where “the public pays to see her and knock on the glass of her jar”. Eighteen months later, instead of being sent back to her people in the Norwegian Sea, she was transferred to Loro Parque in Tenerife. Between shows, the zoo tries to have her inseminated by one of her fellow animals[11], even though it presents itself as a company that has “rescued more than 42,000 sick, injured or orphaned animals with three missions: “rescue, rehabilitation and release”[12]. Less hypocritically renamed United Parks & Resorts Inc. the parent company:

- generated revenues of $1.7 billion[13];
- employs tens of thousands of people;
- is owned by Hill Path Capital, whose founder, Scott Ross, is chairman of the Board of Directors of SeaWorld.[14]

We can only imagine that these money-obsessed financiers are working, above all, for the well-being of marine wildlife.

In March 2025, *Sciences et avenir* reported that the unfortunate orca’s breeding project was nearing completion[15]. What caused her misfortune: her capture is invaluable for a marine zoo. In theory, such a capture is forbidden. But in practice…

*

11. https://www.blog-les-dauphins.com/orque-morgane-maintenue-en-captivite-au-delphinarium-de-harderwijk/
12. https://unitedparks.com/commitment/animal-rescue-rehabilitation/
13. https://s1.q4cdn.com/392447382/files/doc_financials/2024/ar/PRKS-2024-Annual-Report.pdf, p. 48.
14. https://seaworld.com/scott-ross-hill-path/
15. https://fr.news.yahoo.com/enquete-l-histoire-morgan-orque-172554879.html

It's not the first time Loro Parque has been caught up in a wildlife trafficking scandal. Between 1985 and 1990, Tony Sylva, the establishment's head caretaker, smuggled 185 individuals, including hyacinth macaws (*Anodorhynchus hyacinthinus*), for a total of 1.3 million dollars. He was sentenced to seven years' imprisonment[16]. Yet the Loro Parque Foundation claims to work for the protection of the world's parrots, while at the same time selling several hundred of them every year—look for the mistake.[17]

This unfortunate case illustrates two trends. On the one hand, as soon as a species is interesting in terms of return on investment, the smallest individual supposedly treated for release can end up in a zoo after passing through a care center. On the other hand, while some forms of wildlife trafficking still prefer to remain discreet, others, on the contrary, are practised in broad daylight, with the endorsement and even the encouragement of the authorities and administrations in place.

Zoos' interest in care centers would not only make their image a little greener, but would also enable them to obtain supplies directly from the wild, without going through the intermediary of care centers, with the animals passing from producer (their care center) to consumer (their zoo) without any other intermediary.

For me, this is a frightening realization. As a defender of wild animals, I thought I was on the side of the good guys. I'm discovering that the bad guys are *also* on the side of the so-called good guys. Some care centers are rotten. I really became aware of this at the end of 2014, when the French Union of Wildlife Rescue Centers (UFCS) was looking for a site to host its general meeting. I was proud to put myself forward as a candidate on behalf of Alca Torda, and even prouder when I learned that the center where I was working

16. https://www.sauvage.bzh/sommaire/?dir=Chapitre10/Tony%20Sylva
17. https://avesdojorge.com/wp-content/uploads/2021/03/LPF-28-05-2025.pdf

at the time had been chosen as the venue. At the time, 90% of the member centers were unknown to me. I imagined their representatives as passionate, willing, hard-working people who had faced a multitude of material difficulties in general, and financial ones in particular. In short, people who were sometimes forced to tinker reluctantly, but who were always concerned to evaluate themselves, to train themselves and to keep at the center the obsession of our profession: to treat in order to release.

And now, during the AGM, the manager of a care center in Millau, Larzac, explains that he has been involved for years in a program to reintroduce the Lesser Kestrel (*Falco naumanni*), a small migratory Mediterranean bird of prey that is extremely rare in France. The initial flock was made up of injured wild birds that could not be released. In the beginning, reproduction and reintroduction were carried out totally illegally. Regularization took place *after the fact*, but the program worked so well that reintroduction was stopped by order of the French Ministry of Ecology. And, all proud, the guy announces that, from now on, "the young that are born are sold to zoos to pay the price of the seeds".

His words make my blood boil. Being the host of the event, I can't afford to explode. At the same time, I can't afford not to react either. I've just heard of practices that are so contrary to what seems to me to be our common ethic, and above all so contrary to the law, that I take the liberty of intervening. I'm astonished that an animal whose parents come from the wild can be issued with an Intra-Community Certificate (CIC), making its sale legitimate. The speaker reassures me and tells me that there's a guy at the DREAL Occitanie who's been issuing CICs for this type of animal for years. For kestrels, he has awarded—the Holy Grail—CIC source Cs, i.e. for commercial purposes, whereas these documents are exclusively reserved for regularly acquired animals of captive origin. For me, it was a wake-up call. I haven't spent my life fighting against wildlife trafficking only to discover that the guys who are supposed to be of my background and persuasion have, in fact, gone over to the dark side of the force.

The general meeting ends with a terrible *clash.* The party was more than spoiled. Gilles Moyne and I decided to withdraw our centers from the UFCS and resign so as not to be associated with what turned out to be a sinister farce. My pride was replaced by a mixture of dismay, bitterness and anger. All of which has fueled my determination to pursue what I know, in my heart of hearts, to be both:

- my battle;
- my mission;
- and my raison d'être.

*

My experience at the UFCS General Assembly made me realize that not only do wildlife care centers regularly supply zoos, but above all that this practice is carried out with a rare lack of complexes.

For example, the Millau center, which is very active in the field of supplying living things to the zoological world, announced via its Facebook network in 2017 that it was exporting a griffon vulture to the Rocher des Aigles, under the pretext of being "impregnated"[18]. The story doesn't say whether the Rocher des Aigles will make a donation to the Millau center to thank it, as some zoos do.

What we do know, however, is that the Rocher des Aigles had its own care center, which operated for 37 years. Thirty-seven years during which the establishment claims to have taken in and cared

18. https://www.sauvage.bzh/sommaire/?file=Chapitre10/Rapacecentrede-soins/13.png

for 4,000 birds, while unscrupulously specifying that animals that could not be released remained at Rocher des Aigles.[19]

Some zoos openly shop at French care centers, such as the Doué-la-Fontaine zoo, which in the same year sought vultures for its new attraction “Les Fantômes de l’Himalaya”. In a letter sent to care centers, the zoo’s director informs them that they are ready to receive disabled vultures in their new enclosure, which has a large permanent capacity[20]. You only have to cross-reference this information with the Facebook pages of certain care centers such as Hegalaldia in the Pyrenees to see that birds give rise to disguised financial transactions.

Thus, on May 24, 2016, Hegalaldia announced, via its Facebook page, that it had entrusted the Doué-la-Fontaine zoo with three vultures (two fawns and one Egyptian vulture) disabled and therefore unable to be released.[21]

On January 15, 2017, again via Facebook, the center warmly thanked the Doué zoo for its donation of... €3,000[22]. The operation will be repeated on June 15, 2017 for an additional griffon vulture[23]. The center will thus “export” seven griffon vultures *(Gyps fulfus)* and two Percnopterus vultures (*Neophron percnopterus*) to two different zoos, Doué-la-Fontaine and the Parc des Pyrénées, the latter having also made a donation of €2,000[24].

---

19. https://www.sauvage.bzh/sommaire/?file=Chapitre10/Rapacecentrede-soins/14.png
20. https://www.sauvage.bzh/sommaire/?file=Chapitre10/Rapacecentrede-soins/15.png
21. https://www.sauvage.bzh/sommaire/?file=Chapitre10/Rapacecentrede-soins/hegalaldia2.png
22. https://www.sauvage.bzh/sommaire/?file=Chapitre10/Rapacecentrede-soins/Hegaladial3.png
23. https://www.sauvage.bzh/sommaire/?file=Chapitre10/Rapacecentrede-soins/Hegalaldia4.png
24. https://www.sauvage.bzh/sommaire/?file=Chapitre10/Rapacecentrede-soins/Hegalaldi1.png

In return, the Hegalaldia center provides a communication service via its press organs or directly on site, showing just how "involved" these zoos are in conservation[25].

Of the 25 vultures presented for "Les Fantômes de l'Himalaya", 24 came directly from the wild via French or Spanish care centers. This mix of genres and the risk of it getting out of hand is dangerous, especially in view of the history of care centers in general and the Hegalaldia center in particular, whose regular alerts on its Facebook account and in the media highlight chronic funding problems and a permanent risk of closure due to lack of resources. But the operation seems tacitly validated by the all-powerful LPO, as Allain Bougrain Dubourg himself is photographed in the presence of the freshly recovered vultures presented at Doué-la-Fontaine zoo.[26]

The LPO is no exception to this practice, since its Aquitaine section places fawns in zoos, even though they are not impregnated, directly with the animal exchange of the AFdPZ, the zoo union.[27]

Fawns, like foxes, are one of those species that are almost 100% condemned to captivity when they are taken into care, either because they have been impregnated as a result of poorly managed bottle-feeding, or because their caretakers simply refuse to release them into the wild on the pretext that hunters might kill them. On the other hand, taking in these species generates sales, buzz, Facebook postings and donations.[28]

---

25. https://www.sauvage.bzh/sommaire/?file=Chapitre10/Rapacecentrede-soins/Vautoursladepeche.fr.pdf
26. https://www.sauvage.bzh/sommaire/?file=Chapitre10/Rapacecentrede-soins/RefugeLPODoue.pdf
27. https://www.sauvage.bzh/sommaire/?file=Chapitre10/Rapacecentrede-soins/LPO%3AAFDPZ.png.
28. https://www.sauvage.bzh/sommaire/?dir=Chapitre10/Faon

11.
The Nightmare of Poisoned Macaws

In 2014, just as the Alca Torda center was proving its usefulness as a repository for seized birds, the Manche environmental police informed me that a major seizure had taken place at the home of one of France's largest parrot breeders. In the absence of a *suitable* facility, the parrots were kept by the offenders pending a final decision by the court. In reality, the initial proceedings are two years old, the appeal is due to be heard very soon, and the authorities have good reason to believe that the seizure will be upheld and turned into confiscation. Hence the need to find a home for the birds, some of which belong to very rare species. Clearly, we are a serious candidate to meet the demand, but above all we are the only one. The rest of the process confirmed my contact's premonition. A few months after these initial contacts, I was asked to collect the 70 parrots in question.

An operation like this has to be prepared with the utmost care. Special transport crates had to be made by us, as they didn't exist on the market. And, never knowing what the working conditions are going to be like, a set of intervention outfits, masks, gloves and foot baths are also taken on board. We are accompanied by the gendarmerie and the environmental police. Even if I know this kind of work by heart, I know that I'm going to be scrutinized and that, for the officer in charge of the case, this seizure is carried out under great pressure. When I arrive with a team of three, I can see that the farmer is waiting for us with bated breath. I look at the general condition of the buildings, I take into account the sanitary aspects, I check the size of the doors; I evaluate the size of the rooms to see if we'll be able to get in with one or several people; I estimate how easy it will be for us to evacuate the crates; I list the potential difficulties; in short, I appraise the site.

As is often the case in parrot breeding buildings, these consist of a large central corridor which distributes the aviaries to the right and left. I cross the corridor to estimate the number of birds I'm actually going to collect. Curiously, they've been gathered in the same part of the kennel, as if to limit access to the rest of the structure. I soon discover that a bird which, a few seconds earlier, was in perfect health, is lying on the ground in convulsions. It's shaking, screaming, flapping its wings. It's clearly in agony. The tragedy is that this is a very rare bird, whose heritage value is almost priceless, and whose population in the wild is less than ten to fifteen thousand individuals worldwide. What's more, it's very expensive, and with good reason (we're talking about a specimen costing fifteen thousand euros). The bird in question is a Hyacinth Macaw (*Anodorhynchus hyacinthinus*), not only a magnificent parrot, but also the largest parrot in the world.

I rush into the aviary and swaddle the bird in a terrycloth towel to prevent it from convulsing and injuring itself. Forty seconds later, he died in my arms. I don't understand what could have caused its death.

It's the officer in charge of the control who sheds some light on the matter. He informs me that, while I was shaking the macaw, the owner, in the corner of the door, was showing a small satisfied grin. For the official, the meaning of this mimic was unambiguous: "Hey

hey, since we're not going to have them anymore, you're not going to have them either!" The poisoning of the bird will be proven later. Beyond the revolting stupidity of the act, it constitutes an unacceptable crime against biodiversity.

The tour of the aviaries brings more surprises of the same kind. A few minutes later, a scarlet macaw dies in front of me, with the same clinical picture; and the carnage continues shortly afterwards. Enough is enough. I call a halt to the inspection and ask the gendarmerie to refer the matter to the public prosecutor, explaining that, on the one hand, we're dealing with extremely precious birds and, on the other, of the 70 individuals I have to recover, only 20 are still alive, the others having died before the inspection. When I wonder at their absence from the aviaries, bags of frozen corpses are brought out, containing specimens even rarer than the Hyacinth Macaw. The corpses show signs of mistreatment. Some have been frozen in a state of putrefaction. In other words, they were left in the aviaries for five or six days after death.

The prosecutor agreed with my request to look into the matter. He instructs me to accompany the environmental police to redo the entire inspection. He encouraged me to ask for the seizure of any birds I considered to be in danger in this establishment. What began as a simple seizure, for which a few hours would suffice, turns into a three-day inspection of three hundred parrots. We're changing gear. Mr. Rictus will remember this.

*

In the aviaries, we go from nightmare to nightmare. Sanitary conditions are so deplorable that, when we enter a nursery saturated with ammonia odors, we are forced to use cartridge masks. This inspection gives me the opportunity to give a crash course in environmental policing, including ringing. While the untrained eye might find them normal, I show the officials that this is not the case. The ring is not proof that the bird is legal, far from it—although, often, due to a lack of technical training and knowledge of how to handle the animal, even the most scrupulous officers miss the trick. In this case, most of the rings we come across are the wrong diameter, can be removed, are partly erased, are Belgian when the bird is supposed to be of French origin, etc. These tricks, which are often unknown to law enforcement officers, are just a few examples among the host of shenanigans I listed back in 2016 in a document of over a hundred pages, entitled *Bird trade in France, the reasons for trafficking*. In this file, produced and distributed free of charge to French control officers, I describe the state of trafficking, the legislative context, fraud techniques, laundering schemes and numerous methods that enable officers to detect a fraudulent holding or transaction.

After seventy hours, we leave with one hundred parrots, seized either for illegal detention or detention in conditions that put them in danger. For the duration of the procedure, they are entrusted to the Alca Torda center. During the inspection, however, the director of the Landes hunting federation sent me a message, expressing his surprise at my tardiness. As soon as I can, I reply with a photo of the corpses of the birds we should have housed, lined up in the grass for the count. A snapshot is often better than a long speech! Just as quickly, my employer republished the image on social networks, where it exploded all the counters, surpassing six million views in less than two hours.

The next morning, a torrent of insults poured in on the networks directed at me. The threats that accompanied this outpouring of hatred forced us to seal off the establishment and place it under video protection. The object of this vindictiveness? If we hadn't found the dead animals, we would have euthanized them on site. Faced with our stunned firmness and our decision not to let the defamatory accusations go unchallenged, most critics will eat their hats and retreat, before our promises of legal action kick in. Animal protection is an inflammable subject. It conveys a great deal of emotion and therefore often reveals stupidity or malice, especially when rumors of large-scale euthanasia are blown by breeders in solidarity with their fellow breeders.

Even if this sad seizure is only the tip of a huge iceberg, wild bird trafficking is not inevitable. Like much trafficking, it is the illegal offshoot of a legal trade, hence the importance of clear regulations that make it possible to distinguish unambiguously between what is legal (legal trade) and what is criminal (trafficking). While it does meet a demand, it persists above all because there is no will to eradicate it. Farms of several hundred birds can exist illegally thanks to a combination of two phenomena: firstly, controls are extremely rare; secondly, when controls do exist, they are often of a very inadequate quality and meticulousness. If, tomorrow, an army of duly trained inspectors were sent to France's zoos and breeders, at least a third of all birds would be seized. The problem is that we lack the will, the resources, the skills and the independence of control bodies and reception centers to meet the challenge.

12.
The Impossible Reintroduction

The digital noise generated by the mass grave we discovered had a positive effect: I came under the radar of the World Parrot Trust (WPT), an American psittacine conservation organization. At the end of 2016, a French-speaking WPT representative contacted me to inform me that dozens of Gabon grey parrots *(Psittacus erithacus)*, part of a batch of over three hundred and fifty animals, had been seized for almost three years and placed in a Bulgarian zoo, awaiting a solution. First problem: the zoo doesn't want to keep them. Second problem: I left the Alca Torda center about six months ago. So I have no structure to keep these animals—"around thirty", says my contact. Nevertheless, I find myself cogitating intensely. On the one hand, I don't want to say no, and I don't like the idea of these birds being in danger; on the other, I can't bring myself to cobble together a shaky solution. That's when Noëlle, a parrot breeder who's become a friend, says to me: "I've got a bit of space in my aviaries, and I've got a big empty one that would be perfectly capable of taking in these birds". I jumped at the chance and, some time later, we were off to Toulouse airport to receive the thirty birds from Bulgaria. Curiously, even though they were protected birds, no documents were required by customs, perhaps because this was within the European Community. The transport company doesn't ask us any further questions. In case we had any doubts, this operation shows just how easy it is to move large shipments of this type of bird. The fact remains, however, that when we open the transport boxes, it's a cold shower. After several years in captivity, parrots are in an appalling state. Nearly 90% are unable to fly. Feathers, both tail and wing, are cropped, broken and damaged. I just can't believe it.

It's not uncommon for animals to arrive in this condition when they've just been seized in their country of origin. On the African continent, they are captured with nets, then their wings are cut off with machetes, so that their feathers are ripped out under these acts of torture. On the other hand, it's incomprehensible that birds held in Europe for several years should be in such poor condition. After the astonishment, we take charge of the grays, aware of the challenge and difficulty of the mission. I think of the near-dying seals we once saved: why not the greys?

My first instinct is to try and understand what could have happened to my new hosts. The animal's condition often says a lot about its history. By recording, weighing, photographing and examining the greys, I understand what has happened. When the birds are quietly in their aviary, they always position themselves high up on perches or grills. As they are wild animals, as soon as someone enters the aviary, the slightest noise, vibration or door slam causes panic. Unable to fly, they fall. Examination of each captive confirms my intuition. When we check in our captives one by one, we find numerous leg fractures and fractures of the brevis, the extension of the sternum that forms part of the thoracic cage (where the muscles needed for flight are inserted). The condition of these birds also speaks volumes about the incompetence of the institutions to which they have been entrusted.

My friend and I set out to restore the birds' ability to fly as quickly as possible. When we approach the aviary, we talk, we whistle, we tap very lightly on the door to avoid the surprise effect. In this way, all the birds on the ground have time to position themselves safely, and there's no element of surprise. What's more, the bowls are positioned as far away as possible from the perches, so that we can access them without triggering panic and therefore avoiding falls. It will take time for our protégés to regain their splendor. Nevertheless, within a few days, our protocol is bearing fruit, and the newcomers are regaining a certain serenity, an essential prerequisite for their rehabilitation.

*

Satisfied with our work, the WPT asked us again to take charge of 50 parrots from the same seizure, but held in a Hungarian zoo. We hesitated, as 80 individuals would be a lot. Nevertheless, from a strictly technical point of view, a large group of birds, if its members are more or less in the same condition, and therefore subject to an identical protocol, is worth a try. Noëlle, a friend of mine who breeds birds, offers to make a plot of land near the farm available to the Biome association—which I chair and which is responsible for these operations. The idea is to install mobile homes and build aviaries to group the birds. 65 years old, and a solution to all Noëlle's problems. So be it. We sign an agreement with the Hungarian government (the WPT acts as intermediary and translator) and begin construction of the seventy aviary panels needed to build a large flying area for these 80 birds, probably the largest group of greys held in a single lot. The aim is to refurbish the birds and return them to the wild.

As, until now, everything has been too simple, we learn that these birds will not come to us. We'll have to go and get them from Hungary. No need to get my hopes up! I hire a car and drive across part of Europe, greeted by snow and rain. The rendezvous with the zoo, which has travelled part of the route to meet us, takes place in a truck park halfway along the route. The scenery is lunar. I find myself in the middle of an immense agricultural plain, battered by winds and snowflakes, with, in the center, a restaurant established on the side of the road that crosses the country. We change vehicles, and the return journey confirms just how easy it is to cross Europe with fifty protected birds: not a single control, not a single check, not a single customs officer. It's astonishing, even though I have of course declared this transport to the DREAL, the OFB and the local prefecture.

For the next two years, we're going to be carrying out more and more veterinary assessments, looking after the birds' health, and working hard to get these two groups back into the air. Every day, we have to collect fruit and vegetables from local supermarkets for the 25 kilos of fresh food they need every day. In less than eight months, the birds moult. Broken feathers fall out. Little by little, the parrots regain their vigor. They begin to fly again. We'll soon be able to release them. I'm happy and satisfied with this development. However, for the purposes of this book, we have to ask ourselves what lesson we can draw from the experience at this stage.

The main lesson is that, with few exceptions, zookeepers, however well-intentioned, are not up to the task. The resources devoted by zoos to conservation are calamitous, but the training of their employees is often woefully inadequate. As proof of this, in just a few months we have achieved what structures with far greater resources than ours have not been able to achieve in over three years. But these results are not an end in themselves: we still have to close the loop by releasing our residents.

*

For an animal to be released, it must meet several criteria:

- not be sick;
- it must not show redhibitory behavior (indicating, for example, that it is impregnated);
- not have been hybridized with another species or subspecies;
- be hosted in a territory free from any direct cause of extinction.

We know that these parrots are all of African origin, and that none of them have been bred. In the course of our examinations, we have proved that the parrots carry no pathology, temporary or permanent. So we set out to find a home for the parrots, and that's where the problems begin.

The WPT wants to find a base in East Africa because English is spoken there. However, the natural geographic zone of the Gabon Grey, as its name suggests, is West Africa, a French-speaking area. As a result, the specialized associations I suggest are systematically rejected by the WPT. And then there are the financial problems, since what was supposed to be a provisional arrangement has lasted, and every time I ask for a financial contribution, the WPT replies that the birds don't belong to them, and that we've undertaken to look after them until they're back in the wild, period. I'm beginning to understand better the attitude of the two zoos if the animals were entrusted to them under the same conditions.

Until we find common ground, we continue to care for our residents. We're teaching them to feed on whole, uncut fruit. We're keeping the necessary distance to preserve their wild behavior. We can't wait to see them in the wild again, but releasing them is a major operation. That we take the time to do it right is not shocking—on the contrary. We need to find:

- reliable partners;
- suitable infrastructures;
- favorable contacts both at CITES and with local governments, etc.

After months of procrastination, the WPT announces that it has found a former breeding farm in England where the birds will be placed until they are returned to the wild. A new disagreement arose over the organization of the transfer. The WPT wants to transport the birds in cat boxes, which would jeopardize more than two years' work. During transport, the parrots must not be able to turn over in their transport box or cling to any wire mesh in order to preserve the integrity of their plumage. I'm flabbergasted by the incompetence of the people we deal with, and determined to make sure that our work isn't wiped out with the stroke of a pen by the very people who asked us to do it.

During this period of uncertainty, I embark on an adventure that we'll be talking about shortly: Rewild, a former zoo acquired by a coalition of allies with the aim of closing it to the public and turning it into a reception center for seized animals. So I transfer the birds to the new facility in Brittany.

13.
Hope for Another Zoo

One day in 2008, my best friend, who works at the Pont-Scorff zoo in Morbihan, suggested I get in touch with the owner of Amazonia, a small animal garden on the Ile de Ré. The 4,000 m^2 property features a shell museum, stuffed animals, a dwelling house and 90 parrots, macaws and cockatoos, housed in small alcoves in faux stone caves, arranged around faux islands. The new owner has just bought this rather strange place from its founder. He's not just anyone: at the Carquefou school, the man trains animal caretakers. His problem is that he doesn't hold a certificate of competence, so the prefecture is considering closing his establishment.

As a specialist in hopeless causes, I agreed to take charge of Amazonia. The adventure was short-lived: I had underestimated the owner's penchant for the bottle and its deleterious consequences on the management of the establishment. However, I have time to hire two animal caretakers to restore both the birds and the facilities. The task is made all the more urgent by the onset of winter. As I analyze the accounts, I realize the considerable potential this small business, located in a strategic location, could have had if it had been properly managed. More than a regret, it's a revelation: money from visits could finance a conservation mission!

With the caretakers we've recruited, we're imagining a two-part facility. On the one hand, an area for endangered species, to enable their return to the wild; on the other, an area open to the public featuring animals in semi-liberty, where visitors would be invited to look for the animals rather than observe them in tiny enclosures. The idea transports us. We feel as if we've discovered the magic formula that will enable us to reach a milestone in our shared dream: to help preserve biodiversity while respecting the specific characteristics of wild animals. We've even come up with a name for the site. It will be called the Biome.

But there are still a few tiny details to work out: we need to find a site and funding, especially as Amazonia is sinking. With bankruptcy inevitable, I had to put the company into receivership and return to Brittany. The time had come for me to devote myself fully to the Biome project.

*

My search for a site for the Biome immediately turned to the Landes, where the political and land situation seemed more favorable. I moved the dozen or so animals I'd been taking care of at the request of the authorities to trusted adopters, and arrived in the department I now know well, with my goal firmly in mind... but no job. As luck would have it, a veterinary friend of mine wants to turn a practice into a clinic. On the strength of my experience with Alca Torda, UMSOM and my reception center, I offered him my help, and I was hired. For two years, I built his clinic, I negotiated with the laboratories that wanted to come and work at my client's premises, I worked on the site myself... and, at the same time, I found myself helping out at the Alca Torda center. I've been called back, as the structure no longer has any capacity. The emotional blackmail by the president of the hunters' federation ("Alca Torda is your baby, you can't just drop it like that...") worked so well that I once again became a volunteer instructor at the center, compensated with mileage. I soon added another string to the center's bow by providing training in wildlife care. It's the only place in France to offer this service. On the strength of this success, I relaunch my idea of a center to take in seized parrots, explaining to hunters that:

- since they are responsible for protecting nature;
- since they own two unused hectares;
- and since we can apply for European subsidies;

all the planets are aligned for the adventure. And indeed, things are looking up. On the equipment side, with a budget of €300,000 allocated as part of a cross-border project with Spain, we are building very special aviaries and boxes, as parrots require facilities adapted to their lifestyle. On the administrative side, the close ties between the Office National de la Chasse et de la Faune Sauvage (ONCFS) and the Landes hunters' federation are bearing fruit. Our file arrived at the right time in the hands of the right people and, oh miracle, avoided getting bogged down for once under other files and thick layers of dust. An inspection is carried out. It validates the principle of the structure I'm working on. The Alca Torda center, which specializes in wildlife care, officially sets up a new branch to handle seized animals. In less than two years, the center proved its indispensability. It quickly filled to capacity, with record seizures and peaks of almost five hundred animals taken in, testifying to the scale of the problem: I wasn't wrong! But this success had a downside: the friends of the traffickers could not remain indifferent for long...

*

You have to hand it to Jacques W.: he never gives up, and he's a stubborn Breton who admits it! Every time the cause of wildlife advances, he appears and struggles to turn progress into setbacks, especially when it comes to seized animals. One step forward, two steps back, more if possible.

The Alca Torda appendix is no exception. Not that we have reached a spectacular size, capable of making all the traffickers tremble by signalling that the seized animals can actually be taken away from them; but, on our scale, we have demonstrated the usefulness and above all the effectiveness of such a structure. In the meantime, thanks to the Alca Torda center, our agents have a site where they can receive seized birds, enabling them to carry out their procedures and do their work. Our success is a cause for concern. There's a real risk that we'll become an example that will spread and snowball. We have to put a stop to it.

Consequently, when we presented our center to the national authorities, Gérard B., Jacques W.'s subordinate, sent an e-mail to all DDPP agents[29]. He warned those who had received our brochure. According to him, our initiative had not been validated at the highest level (which is untrue), had not been approved by the Ministry (authorization to open by the State services is equivalent to approval ment), and had not followed the Ministry's guideline that seized animals should be sent to non-commercial establishments, in this case: breeding farms and establishments for public presentation. Yes, it's spelled out in full: "breeding farms and establishments for presentation to the public" are non-commercial structures. Of course!

As unfounded and grotesque as it is, the warning shot is a violent one. The president of the Landes hunters' federation had to go all the way to the top of the Ministry to make up for it. We escaped the boycott while, spurred on by adversity, I gave more thought than ever to making the Biome project a reality. I'm delighted to see that

29. https://www.sauvage.bzh/sommaire/?file=Chapitre13/MailBassot.pdf

the idea has won over a large number of people. Yes, commitment to wildlife protection is tough enough to allow me, when the opportunity arises, to crack a satisfied smile, and when I do, my smile is fully justified. The partnership with hunters is solid. Politicians are interested, because in this part of the Landes, it's difficult to attract tourists. A wildlife project can be very appealing. Since the first sketches, the concept has become clearer: the Biome is intended to be a multi-species establishment (birds, monkeys, felines, etc.) for conservatory breeding, i.e. breeding in captivity before returning to the wild.

To help finance our needs, which will contribute to our independence, we are considering, inspired by what we have seen and experienced at Amazonia, that part of the establishment be presented as a zoo. We did so reluctantly, of course, but at the time, in my mind, "zoo" wasn't a dirty word. Admittedly, this lucrative *business* turns wild animals into cash machines. However, in theory, a zoo can also be a place for:

- raising awareness of biodiversity;
- conservation of endangered species;
- preservation of genetic heritage;
- and scientific research.

In theory. At the time, I believe so, because, in principle, we can't rule out the idea that an establishment presenting animals to the public should be guided by commendable ethics and an awareness of the specific needs of wildlife. But there's no question of it being a classic zoo. The zoological part of the Biome will be resolutely atypical. It will be spread over huge areas where the animals will be dispersed. To see them, as we envisaged from the outset, you'll have to settle down with a pair of binoculars, or put your ear to the ground, day or night, at the invitation of specially trained animators. We plan to add to these activities the benefits of virtual reality via oculus to limit the frustration of visitors who might regret not seeing enough animals. The idea is that these substantial investments will enable us to generate the necessary funds to finance our core project.

Today, I've changed my mind. If I were to set up a project of this type again, and why not, I wouldn't for a second consider opening even a small part of the establishment to the public. I've seen all too well how harmful the commercial exploitation of wild animals is for the species they keep. At the time, I still had blinders on. I imagined that, if many zoos mistreat wildlife, it's because of a lack of competence or an excessive concern for profitability. I still believed that wild animals could be kept in good conditions. Since then, I've gone back on my illusion. We can do whatever we like, but wild animals under duress will never have the life they should have. The only exception that might make captivity tolerable, for want of a better word, is for captive animals to reproduce in order to save the next generations. With hindsight, however, I'm almost glad that Biome didn't participate in the economic exploitation of animal confinement... since it remained just an idea!

*

In the space of a few weeks, everything fell apart, but it wasn't a big deal. Around 2017, I had to face the facts: the entente cordiale I had established had come to an end. The hunters' desire to take control of Alca Torda, their pressure to redirect the business under the pretext of competition from neighboring care centers, and the virulent hostility towards me from the federation's president since he discovered that—oh scandal! I was living not with a girl, but with a boy, led to the inevitable. The day we were to sign the purchase of the land to build the Biome in place of the old pheasant breeding center, the president backed out and, without giving the slightest reason, refused to put into practice his promise of an emphyteutic lease, even though it had been signed in the presence of a notary. For seven years, step by step, we had put together a project in a collegial and concerted manner, spending hundreds of hours in meetings, mobilizing our energy and skills to the point of losing our breath. We had also managed to secure €850,000 in private partnerships, in addition to bank loans and a guarantee from the public investment bank (BPI). This gigantic amount of work, human investment and hope vanished with a snap of the fingers. By stepping aside, the president of the hunters' association led our partners to withdraw, in a domino effect: everyone goes if everyone goes; if someone withdraws, no one goes, because the project no longer holds together. And if there's no place left to build the Biome, that's the end of the project.

The man who ruined the project has never agreed to explain himself, but I think I've pieced together part of his rai sonnement. When the president of the Landes hunters' federation realized that he would never be able to get his hands on Alca Torda, the atmosphere deteriorated and quickly turned explosive. The context accelerated the process. At that time, an avian flu epidemic was raging in the region. For me, this was a major concern. At the center, I keep a few native birds, but I also keep around three hundred and fifty parrots from criminal proceedings—so they have the status of exhibits. The regulations are unambiguous: if I bring in a European bird carrying avian flu and the individual contaminates exotic birds, I'm obliged to euthanize the entire flock, including some extremely rare birds. As a result, I close the center to indigenous animals. It's my duty and my authority as a manager, but the president is furious.

When I realize that, taking advantage of one of my rare absences, he's trying to get around my refusal, I understand that my work is becoming impossible and that my mission, which is to act for the protection of wildlife, is in jeopardy. I immediately resigned from the Alca Torda center, which found itself without a manager and no longer allowed to operate. The Landes experience thus ended very badly... without really being over, since I initiated legal proceedings against my former partner—proceedings which are still in progress at the time of writing. This abortive experience confirms that what often kills projects is the pursuit of self-interest. The hunters had an interest in developing Alca Torda: they were confirming their role as defenders of nature; and, above all, they thought they could control the business to their heart's content. Not so. In this kind of ship, the owner is indeed the hunters' federation, but the captain is the only master on board; and the captain happens to be the owner. Me. When President realized this, he became furious enough to trash the place. To be angry, to blackmail, that's his right; as to whether to ransack everything was also his right, the law should soon decide. In the meantime, with or without Biome (without, that is), I'm not thinking for a second of throwing in the towel and remain, in the words of the referee at the start of an MMA fight, *ready to fight*!

14.
The Worm is in the Fruit

Fighting wildlife trafficking sometimes involves risks I hadn't previously considered. At this stage of my career, the various establishments I've set up, the expert reports I'm often called upon to provide and the training I give to the environmental police make me, *at the very least,* a credible, almost referential figure in the field. But since nothing can be taken for granted, and people's interests can change according to their desires and needs, in 2017 I went from being an expert in the field to a pariah to whom no one has the right to appeal, by order of the environmental police. How is this possible, you may ask? Well, it's very simple.

On February 2, 2017, I was summoned to the gendarmerie in the village where I live to be interviewed by the Landes DDCSPP about the sale of birds by Régis H., the director of the hunters' federation. I was presented with three documents, dated two years previously, for which I had to justify the presence of my signature.

The complexity of the exercise is significant for me, as I'm being asked to call on my memory, which, at 55, I still don't fully understand. I'm one of those people who never forgets a smell, an image, a piece of music or a photo, who can store an incredible amount of information as long as it's based on experience and not theory, but I'm often unable, without special concentration, to remember what I ate or watched on TV the day before, a face I saw an hour earlier or a first name given three minutes earlier.

Under these conditions, how can I spontaneously answer a question about a document I signed... two years earlier? It's impossible to use my memory, so I concentrate on the content of the documents, trying to remember the context in which I might have drawn them up and signed them. The exercise is made all the more complicated by the fact that I spent six years at the head of the Alca Torda center, including five years as manager of seized animals. I signed hundreds of administrative documents relating to the seizure, placement, branding and adoption of birds.

The three documents in question are attestations of marking, i.e. forms filled in by the person who either bands or chips a bird for identification purposes.

This document must be completed in a highly standardized manner and, in the case of microchipping, must be co-signed by the veterinarian performing the procedure. Of the three documents, two concern the fitting of microchips and one the fitting of a ring. After careful reading, I realize that I would never have completed such documents in this way. This aroused suspicion in my mind, and I became intrigued by the very structure of the document. It's a CERFA, a form created by the State, with a reference number, and I find that the various tables and columns that make it up are not aligned, making me think of a montage.

It's only when you look at the signatures, mine and the vet's, that the doubt disappears: these documents are forgeries. In fact, our signatures are positioned in exactly the same way, and it's only when I superimpose the three documents and hold them up to the light that I realize that they are photocopies.

The problem is that the birds in question come from a pair loaned to the director *of* the Landes hunters' federation, who is also in charge of the administrative management of the Alca Torda center. There is every reason to believe that he is guilty, *at the very least,* of the offence of illegally marketing protected species. So the music changes completely. At the end of my hearing, I go into another room to lodge a complaint against the said director. I realize that this complaint is going to be a bombshell, and so much the better. The owner of the only French establishment authorized to manage seized animals:

- draws on "stocks" of seized birds;
- takes advantage of his status to sell protected seized birds;
- and betrays a former manager by forging a document.

It's a mess, and I intend to make it known.

I had to wait six months to be interviewed for the first time by the Mont-de-Marsan Gendarmerie's research brigade, by an investigator who knew the respondent. I left the hearing satisfied, thinking that the case was entering its active phase. I would spend another six months without any news of the proceedings, until that morning in November 2017 when I was summoned for a final interrogation with a view to closing the case, the gendarme in charge of my case told me by telephone the day before. So it was with complete peace of mind that I made my way to the Mont-de-Marsan gendarmerie. As soon as I arrived, I was informed that as of this minute (8:30 a.m.), I'd been placed in police custody for breach of trust, trafficking in wild animals, illegal practice of veterinary medicine, complicity in forgery and I don't know what else, including collusion with the administration, but those are the unofficial charges.

It didn't take me long to realize not only the seriousness of the charges against me, but also the injustice of such accusations in the light of the three decades I've spent fighting, often on a voluntary basis, against animal trafficking and for the protection of animals. I will also soon realize that the die is cast, and that the powerful Landes hunters' federation has succeeded in obtaining a rescue operation from the environmental police. The stakes for hunters are indeed enormous, and the complaint lodged against the federation's director risks shaking the foundations of the entire institution.

*

To understand this, we need only look at the facts, which suddenly fit together like a jigsaw puzzle around a central player: the environmental police, otherwise known as the ONCFS for Office national de la chasse et de la faune sauvage. At the time, its Board of Directors comprised twenty-two members[30]:

- four representatives of government ministries, including Agriculture, Environment, Budget and the Office national des forêts (ONF);
- one representative of rural landowners' organizations;
- two representatives of nature protection associations;
- two representatives of ONCFS staff;
- two representatives of pro-hunting agricultural organizations;
- ... and eleven representatives of the hunting community.

In other words, more than 60% of the institution and its 70 million euro budget (rising to 74.7 million in 2024) are controlled by hunters, headed by Henry Sabarot, President of the powerful Gironde hunters' federation, who is also a regional councillor[31], a fervent supporter of ortolan hunting and a candidate for the position of President of the Médoc Regional Nature Park. He's an intimate of the president of the Landes hunting federation, which explains why. Add to this the fact that:

- the ONCFS police director was at the time the former chief of staff of the Landes prefecture, at a time when the same

30. https://www.sauvage.bzh/sommaire/?file=Chapitre14/CAOncfs.pdf
31. https://www.sauvage.bzh/sommaire/?file=Chapitre14/SabroCR.png

prefecture was illegally granting safe-conducts to ortolan hunters, despite the bird's protection and the ban on hunting it[32]. This derogation brought the Landes prefect to court[33];

- the recently retired Director General of the ONCFS was a close friend of hunters, particularly from the Landes and Gironde regions. Their "friendship" is visible even in the press, which shows them together on hunting trips, proof of their closeness[34];
- that the most recent chairman of the ONCFS board of directors has Vuitton as his surname[35]... Si, si, the biggest tanner of exotic skins in Europe.

Doesn't that sound like a bad-smelling wind?

In a scenario worthy of a bad B-movie, the unscrupulous environmental police are sent in to rescue the director of the powerful Landes Hunting Federation. In no time at all, the DDCSPP officer is questioned for over three hours, then dismissed from the case, and a gagging procedure is set up against me.

On the day of my interrogation, I was confronted by Tic and Tac in the role of the shock duo. The thesis they were to develop over a twelve-hour period revolved around a hypothetical collusion on my part with the veterinary services, with a view to framing the director of the Landes hunters. The complaint I lodged is hardly mentioned at all. On the other hand, there's a lot of talk about my six years of managing Alca Torda. Oh yes! If you scratch a bit, you might find a mistake in my work. Oh, even the tiniest error would

32. https://www.sudouest.fr/landes/audon/chasse-a-l-ortolan-dans-les-landes-la-preuve-qu-il-y-a-bien-eu-une-tolerance-2983927.php
33. https://www.sudouest.fr/landes/dax/landes-un-ancien-prefet-a-la-barre-pour-parler-ortolan-3215710.php
34. https://www.sudouest.fr/environnement/chasse/gironde-les-chasseurs-n-ont-pas-vire-nicolas-hulot-2974238.php
35. https://www.lanouvellerepublique.fr/loir-et-cher/commune/vernou-en-sologne/hubert-louis-vuitton-president-de-l-oncfs

do the trick and make me lose all credibility. Scratch, my friends, scratch, I'm very serene! In fact, as the hearing progressed, Tic ou Tac finally told me that the director had admitted selling birds that had been banned from the market, and that he had made fakes. Contrary to appearances, this is bad news: it's becoming essential to push me down so that the hunter doesn't sink alone. I would later discover that, in the judicial records processing system (TAJ), it was specified that the proceedings had been initiated on the basis of judicial information from the ONCFS CITES-Capture brigade, the very people I had trained and assisted for years.

The proceedings remained on the prosecutor's desk for seven years, to the point that I referred the matter to the examining magistrate to dig up the file from the pit where it was sleeping. Because let's be clear, the aim of this gagging procedure, as its name suggests, was to shut me up. Well, ratoche.

On the other hand, in addition to the administrative fatwa decreed by Jacques W. 30 years ago, it will have resulted in all agents in France being forbidden not only to work with me, but also to refer other investigative services (gendarmerie, customs) to me[36], under the pretext that an investigation is still underway.

No one has any interest in seeing it through to the end, and everyone seems inclined to let this file lie dormant until it buries itself in the meanders of judicial administration. It's a safe bet that, by the time they are brought before the courts, the various parties involved will have been moved, shielding them from any legal or professional consequences, or so they think.

36. https://www.sauvage.bzh/sommaire/?file=Chapitre14/Image-1.jpg

15.
Wildnapping a Bear

Sometimes, the sign of a possible connivance between the administration and trafficking flashes through a simple newspaper article. Thus, in May 2018, *Le Figaro*[37] announced that a wild bear cub had been recovered by the environmental police in the Pyrenees and sent totally illegally to the home of a certain Sébastien M., who is both a wildcat breeder and a bear trainer... and who has no authorization to collect animals from metropolitan wildlife. The risk of leaving a bear cub with a breeder and bear trainer is that, unwittingly or not, it becomes an animal lost to the wild. Behind this scandal, which consists of entrusting an animal to someone who has neither the right, nor the expertise, nor the competence to take it in, is the DREAL Occitanie. I understand that the information, which had long been brought to my attention, was no mere rumor, since the environmental police issued a press release to justify themselves, stating that "the bear was deposited in the establishment in question on the advice of this administration"[38].

37. https://www.lefigaro.fr/actualite-france/ourson-sauve-dans-le-tarn-les-inquietants-rates-des-services-de-l-etat-20190709
38. https://www.sauvage.bzh/sommaire/?file=Chapitre15/null.pdf

The agent in question, David D., "head of the local Washington Convention office" at the DREAL Occitanie, seems to have made a specialty of encouraging trafficking and smugglers. And not for small quantities! No, the enormity of the facts is staggering. A few years earlier, he had made headlines among OFB agents for his support of traffickers, in total breach of his duties as an official of the Ministry of Ecology. In 2016, a very large reptile importer based in Toulouse, les établissements Savanna, was implicated by a trainee turned whistleblower Disgusted by the incompetence of his supervisors and the appalling mortality rate he had observed, he posted a video on his social networks that led the establishment straight to court[39]. At the hearing, an important witness came to the defendant's rescue, and criticized the procedure carried out by a local environmental police officer. The witness in question was David D.[40]

This is undoubtedly a textbook case in France, which has triggered an outbreak of hives among control officers, without anyone doing anything about it internally. The notorious DREAL agent took advantage of the situation to try and justify to the court the unwarranted and obviously complacent issuing of CICs.

The Savanna manager was nevertheless sentenced to a heavy fine and a 10-year ban on working with animals, which is no mean punishment for France.

Curious by nature, and accustomed to off-the-wall civil servants, I've long had a keen ear for the slightest piece of information that came my way from Occitanie. And to be honest, I've done quite a bit of scratching on the subject. The reputation of this official has spread far beyond the borders of Occitanie, to the point of interesting a reptile importer from the Paris region.

39. https://www.zinfos974.com/Un-employe-revele-des-pratiques-de-mal-traitance-dans-une-animalerie_a64678.html
40. https://www.sauvage.bzh/sommaire/?file=Chapitre15/processavanha.pdf

La Ferme Tropicale, the importer in question, is to date the biggest of its kind for reptiles in France and perhaps even Europe, with sales of over 2,500,000 euros (2016)[41]. It has two stores in Paris and Lille, and a kennel in Combs-la-Ville. Its manager, Karim D., sits on a number of commissions, including the Commission pour la faune sauvage captive in the "collège des personnalités qualifiées"[42]. He's on every reptile program and show, and his reputation is undoubtedly a source of protection.

Incidentally, he trains environmental police officers in reptile restraint, which seems to earn him a certain administrative complacency in the event of an inspection.

All the CITES permits needed to market the animals must be issued by the Paris region DREAL, known as the DRIEA, since this is where the reptiles arrive on French territory. And yet it's in Toulouse that he has his CICs issued by the DREAL Occitanie. All he had to do was set up the administrative headquarters there to ensure that permits were processed in this "favorable" region.

There are so many applications, given the volume of imports, that the CICs would be processed by an independent design office in Toulouse. So far, nothing forbidden, though. What does make it awkward, however, is that the office in question is run by the girlfriend of David D.'s department head, who is also in charge of the biodiversity division at DREAL Occitanie.

*

The above account certainly denounces the illegal practices of some individuals. More than that, however, it aims to highlight the intertwining compromises that are essential to trafficking.

41. https://www.sauvage.bzh/sommaire/?file=Chapitre15/bilanfermetropical.pdf
42. https://www.legifrance.gouv.fr/jorf/id/JORFTEXT000037833661

More often than not, there is no such thing as the wheat on one side and the chaff on the other. When it comes to wild animals, many players are convinced that making deals, scheming, scheming and cunning are essential to survival, success and prosperity. Everyone (or almost everyone) dreams of holding on to the other, creating their own network and exchanging "services", because it's better to spin in company. In this sense, wildlife trafficking feeds on a rather appalling interdependence between crooks, which is the stuff of organized gangs.

This impunity says something serious and worrying about the structural institutionalization of the trade. Here again, I could have let myself be overwhelmed by the feeling of drowning by trying to be honest, committed and firm, while I'm wading, along with a few others, in an ocean of scoundrels. What I've learned from these various cases has had the opposite effect: there's no way we can leave France's wildlife to such crooks.

16.
Capturing Vinaceous-Breasted Amazons

The fight against wildlife trafficking is not unequivocal. Sometimes the traffickers win. Sometimes, the advocates of legality win. Often, the victories of one side or the other remain relative. Witness the two biggest zoo inspections I took part in in 2017.

Generally speaking, zoos are only occasionally inspected by the authorities. They are simply "visited" on rare occasions. The nuance is important. For a visit, the captivity industry is usually notified in advance. They welcome the agents, show them aviaries and animals from afar, buy them coffee and escort them back to the parking lot. The officials, convinced, make the implacable observation that is expected of them: all is well. In the case of France's largest zoo, the prefect may go so far as to ask the zoo to *schedule* a visit at its convenience. The inspector must comply.

Nevertheless, it sometimes happens that agents from several departments get together, compile and compare data, raise a question, investigate and end up doing the job they've been assigned to do: inspecting. As part of this interdepartmental cooperation, the DREAL (the regional directorate for the environment, food and housing), the OFB (the French Office for Biodiversity) and the OCLAESP (the Central Office for Combating Environmental and

Public Health Violations) have carried out a lengthy investigation into the Cerza zoo (for "Centre d'études et de reproduction zoologique augeron"), near Lisieux, and far too many warning lights are on for them not to intervene.

Covering 80 hectares, the zoo is home to 1,500 animals from 120 species. Cerza is France's third most profitable zoo, with sales in 2024 of 8,400,000 euros and a profit of almost 2 million[43]. A certain Thierry Jardin, a member of a large family specializing in the field, co-founded the site with his brother Patrick in 1986. The siblings own four establishments, two in Normandy for Thierry (Cerza and Biotropica), and two in the Paris region for Patrick (Terre des Singes and Parc des Félins, grouped together since 2024 into a single commercial entity, Lumigny Safari Réserve).

Together, the four establishments generate sales of more than 18 million[44] for a total of almost one million visitors a year. The Cerza alone, for the years 2021 and 2024, has distributed to its shareholders 8.5 million[45] in dividends, as many heavyweights in the sector.

At the end of 2017, I was asked by the public prosecutor to assess the facility. I was seconded to the Gendarmerie to support the environmental police. I spend eight days on site. As you might expect, the inspection is going to be quite extensive and, although I'm a tough nut to crack, I'm feeling a certain amount of pressure. The operation begins with a search. To underline the seriousness of the process, hostilities open in the presence of the public prosecutor and the deputy public prosecutor.

I'd never been to Cerza, although I'd heard a lot about it. I was expecting a high-end establishment. Instead, I discover a farm,

43. https://www.sauvage.bzh/sommaire/?file=Chapitre16/CERZA-compte2024.pdf
44. https://www.sauvage.bzh/sommaire/?dir=Chapitre16/ComptaCERZA
45. *Ibid.*

admittedly home to giraffes, rhinos and felines, but a farm. At first glance, nothing is right:

- pestilential effluent is poured directly into the river;
- the enclosures are cramped, dirty and poorly maintained;
- and the facilities are immediately inadequate.

As is often the case, the inspection is carried out in two parts, the first involving a study of the documents and the second a check of the animals' markings and origins. I'll also be checking that all sanitary standards are respected, as these are so important to me and so much a part of the zoo's obligations. At the gendarmerie's request, I choose the set of documents I want to study. In so doing, they are placing even more responsibility on my shoulders, since they are clearly relying entirely on my expertise. My choice will obviously be all documents relating to the legality of the animals, their provenance and everything to do with trade authorizations. Inevitably, the choice is made on the basis of target species and heritage issues. This selection remains enormous. It corresponds to several thousand paper documents, computer files and e-mails, seized by the gendarmerie to be scanned or copied, under the stunned eye of the zoo's directors. In the process, they made the usual threats, ranging from "you're going to get into trouble" to "I'll pass the information on to the union so that it can refer the matter to the President of the Republic". And strangely enough, this kind of threat works very well and, in less time than it takes to make them, the tension rises a notch among the gendarmerie and by extension for me. I'm always amazed by the threatening attitude of the person being checked when, if he's got nothing to reproach himself for, there's no reason to give way to the threat.

After this moment of stress, the actual inspection begins in the middle of the afternoon.

*

On the second day, the test continues:

- examination of documents;
- visit to the freezers;
- capture of certain animals (which I prefer to take care of, as the zoo staff don't seem to be very competent in this area);
- chip control with telescopic pole for
  - Bengal tigers;
  - Indian rhinoceroses;
  - Malayan tapirs;
  - gibbons...

A search of this type of establishment is never carried out without tangible evidence to present to the public prosecutor. One of these elements concerns the amazon wines, whose administrative control shows that their origin is likely to be dubious.

Four Amazons are present at Cerza, two in a relatively small breeding aviary and two in a huge display aviary. While capturing the former poses no problem, capturing them in the 1,000 m^2, six to eight-metre-high display aviary will prove rather complex.

To achieve this, we ask that the birds be fed only inside the building adjoining the aviary, so that when they go to eat, they are locked in and captured. Very quickly, I realize that our target amazons not only haven't returned, but the vet seems to be implementing a strategy to prevent them from being captured. On several occasions, we give instructions to leave the building doors open and the lights on until late at night, as the Amazons are bound to be attracted to them. And every evening, the building is closed, even though they are less than two meters from the target. This attitude reinforces our determination to catch these birds. On the last evening, the gendarmes gave the order to leave the doors open until we gave the order to close them. Despite the instructions given by the legal representatives, the zoo vet comes by to close the hatches so that the wine amazons we've spotted can't be traced.

In the evening, I debrief the situation with the environmental police officers. One of them turns to me and says:

—Well, if you want, I've got a technique for catching them!

I look at him quizzically.

—When you want to ring a bird in the wild, you catch it with a net, don't you? Why don't we use these nets to catch them in the aviary?

I shake my head:

—But the nets will be dead by the end of the operation. When the bird's inside, you can only get it out by tearing the net.

—Don't worry, we're going, and too bad for the net!

The next day, we arrive with our nets. I equip myself with an on-board camera to protect myself against any complaints about restraints. We observe the route taken by the birds to escape from us. In an enclosed area, a parrot tends to repeat the same route when trying to escape. It was on this principle that I caught the first two Amazons, just by staying put at the point where they passed while three keepers were exhausted chasing them. So we set our nets in a strategic place in the aviary and disturb the birds so that they fly away. The first attempt was a failure. For the second, we moved the nets up slightly and repeated the operation. A quarter of an hour later, both birds were in our hands. It's a great victory for us, but one that makes the vet's stomach turn. The birds are presented to her so that the fleas can be checked in her presence. She is asked to confirm the numbers. Stuck, she complies; and we're not disappointed with our catch.

For one thing, these birds are listed in Appendix A of CITES. As such, they can only be sold with a ring. But they are chipped. Secondly, the birds come from Beauval, which will enable the courts to intervene in a zoo that is usually inaccessible. An agent will be commissioned to carry out an inspection. He will conclude that the

Amazons' parents were bought four years ago in a pet shop—which is illegal—and are not ringed, which is also illegal.

*

Very soon after the inspection, soldier Cerza's defense was organized. Did the public prosecutor's office sense the pressure? In any case, the inspection and the suspicions of trafficking were reported in the press by the legal authorities.[46]

As far as the locals are concerned, there can be no wildlife trafficking in this zoo, an institution that welcomes so many visitors and employs so many people. Caught red-handed, the zoo pleads a simple discrepancy between "the administration's time, which does not correspond to the animal's time and, in some cases, poses compliance problems regarding their identification".

Well, it doesn't mean anything, it's just political rhetoric, but it's obviously acceptable. Some people, including non-specialized magistrates, let themselves be charmed by this imbitable discourse, decorated with a few words like biodiversity and sales. And, like a suppository, it goes down well.[47]

And yet, I'm convinced after reading thousands of seized documents and hundreds of e-mails, which I'll spend six months examining to discover some mind-boggling things. Among them:

- births of animals without parents, returned the same day to the Parc des félins (the brother's zoo);
- animals which, as soon as they enter Cerza, are sent to Biotropica, the other establishment in Normandy (the reason

46. https://france3-regions.franceinfo.fr/normandie/calvados/zoo-cerza-coup-enquete-trafic-animaux-veterinaire-repond-aux-accusations-1852010.html

47. https://www.leparisien.fr/environnement/enquete-au-parc-cerza-pres-de-lisieux-le-zoo-se-defend-de-tout-trafic-d-animaux-10-07-2020-8350593.php

is simple: Cerza has compulsory health clearance to import animals, Biotropica doesn't), which poses a major health risk for animals and visitors alike, since the transfer to Biotropica is exempt from all health controls;
- animals disappearing in staggering numbers;
- euthanasia by convenience;
- illegal issuing of health certificates...

In my report to the public prosecutor, poetry and narrative have no place. The result, however, weighs in at a hefty two hundred pages, squeezed together like an espresso at the bar, and overflows with worrying reports:
- absence of opening authorization;
- non-compliance with sanitary approval;
- failure to implement biosafety measures;
- no entry/exit register;
- 15 non-compliant facilities;
- 12 species of mammals born without spawners on site;
- mortality ranging from 43% to 100% for certain species, to name but a few.

By way of example, the mortality of 30 Sika deer in nine days, 41 wallabies in three days, with the only annotation in the documents: "dead", which clearly designates them as the victims of an unjustified mass euthanasia, at least that's what I suspect. Otherwise, autopsies (obligatory in cases of suspicious death) would have been carried out, or we'd be dealing with an epidemic that absolutely had to be reported to the health authorities. Let's talk about the health authorities. They have approved a quarantine facility, a compulsory import facility, in which a tiger wouldn't even fit. A common garden shed was used as a quarantine facility, but that didn't stop the health authorities from illegally issuing an approval. The animals were not really quarantined, but that's not far off! There's also the case of the ostriches "handed over" to France Autruches, the leading ostrich meat company before it went bankrupt in 1999, which raises the question of the mission claimed by Cerza, supposedly "committed to the protection of wild animals in their natural environment".

*

As this first inspection led to others, a second mission was commissioned. The public prosecutor yielded to pressure from Cerza not to appoint me as expert in this new episode. I regret this all the more as the euthanasia of a banteng, a South-East Asian ox, was scheduled for the day of the inspection. The animal is a magnificent monster, weighing almost eight hundred kilos when fully grown. At the very moment when the public prosecutor, environmental police and gendarmerie raided the facility, the animal was shot dead in a shed... because it was too complicated to manage. This kind of exaction reflects the feeling of impunity that reigns in zoos.

The banteng is classified as endangered by the IUCN, with a population of less than 5,000 individuals in the wild, but at Cerza, euthanasia is carried out on the grounds of collection management. On the zoo's website, you can read about the species: "This year, special in 2020, has seen the Cerza banteng group grow. Two cubs were born in May, and these Asian cows are extremely endangered in the wild. These births are therefore of particular importance for the park, which has long helped to manage the reproduction of this species in zoos"[48]. The proof is in the pudding that conservation is nothing more than a commercial pretext that zoo directors don't give a damn about, as long as the public is there, and the profits along with it.

48. https://www.sauvage.bzh/sommaire/?file=Chapitre16/Banteng.png

For several years, the case continued. Prosecutors change. Today, the consequences for the offending zoo are negligible. Basically, a slap on the wrist, a kiss and off to bed. Cerza even managed to secure the support of former French Minister of the Armed Forces and President of the Normandy Regional Council, Hervé Morin, who declared in the press, in the midst of the judicial investigation: "I want to send a message of confidence and solidarity to Thierry Jardin, a passionate man, who is the subject of abominable proceedings and accusations. I knew all about the growth and extraordinary success of this project", salutes the President of Normandy. "What we have here is a major tourist hub that irrigates the entire region and its hinterland. It's a real jewel."[49]

In the summer of 2023, "a judicial public interest agreement" was ratified by the president of the Lisieux judicial court.

Seven years after the start of the proceedings, and with a further three-year delay, "the CERZA company [...] is contesting its guilt", "the validation order does not entail a declaration of guilt", and "the CJIP is not entered in bulletin no. 1 of the criminal record". On the other hand, "full compliance with these obligations will extinguish the public prosecution". A terribly vain move for what is proudly described as "the first CJIP signed by the Lisieux public prosecutor's office in environmental matters"[50]. Too bad for the animals:

- bleached (i.e. robbed from nature);
- killed to be resurrected with the same chip, but in a different body;

49. https://actu.fr/normandie/hermival-les-vaux_14326/pres-de-lisieux-herve-morin-soutient-le-zoo-de-cerza-des-accusations-abominables_35770838.html
50. https://www.justice.gouv.fr/sites/default/files/2023-09/CJIP_CERZA_OV_20230906.pdf

- regenerated without a care in the world (as in the case of this male lynx who escaped and then returned as a female lynx six months later...), etc.

What we have to conclude from this control, which started off under the best aus pices before the soufflet fell back, is that wildlife trafficking is not taken seriously by our institutions. When you try to make an elected official aware of the environmental tragedy these exactions entail, he or she talks about attendance figures at the zoo in his or her constituency, direct and indirect jobs, and therefore reduces all infringements to silly "administrative problems". Is any official trying to show him that these silly "administrative problems" are not technical problems of banding, or nonetheless not insignificant questions of animal welfare, but serious challenges to texts that have the force of law in France and international conventions? This poor civil servant will be crushed by:

- the economic machine;
- the power of zoos;
- and the cowardice of many institutions.

Money often makes the law.

17.
Featherweight

At the end of January 2018, a week after the Cerza inspection, I was again requisitioned by the public prosecutor's office at , accompanied by around twenty gendarmes and agents from the French Biodiversity Office. This time, the target was an establishment whose name I cannot reveal, as the case is still being investigated by the courts. For the sake of the story, I'll call it the Parc de la Plume.

The structure was recently bought from its founder by a couple. Five years later, the new owners will boast that they have made it the second-largest tourist destination in the area, just "behind Center Parcs". As for the seller, who sensed the wind turning and the pigeon showing the tip of its beak, he is a co-founder of UFCS, the network of care centers, and an eminent ornithologist whose work is praised by the LPO in its *L'oiseau magazine*. But above all, he is a man convicted of trafficking wild animals.

So it's a totally off-the-books establishment, as we say in our jargon, with virtually all illegal birds that is the subject of this inspection.

The illegality has been assumed by the seller, since when the new purchaser bought the park and its birds, the notarial deed stipulated "the lack of transparency", we learn in the local press. In other words, I'm buying a space whose main value lies in animals that I know, from the outset, are at least partly illegal. But this quirk is only part of the problem! What the new purchasers certainly didn't know is that the establishment they bought is in fact a money-laundering weapon that will no doubt remain a textbook case in France. It combines a care center for the cause, a breeding facility for prestige and a zoo for money.

Birds taken in at the care center and deemed unsuitable for release were fraudulently integrated into the breeding center to end up in the zoo, all the while obtaining the CIC issued by an agent of the DREAL Normandie in a totally unwarranted manner. When you consider that the official who facilitated these transactions is today a departmental head of the OFB, you can only dream. It wasn't until his successor began to put some order into things, and requested an OFB inspection, that the cat was out of the bag.

The founder of Parc de la Plume even invented a formula, which he left in the park's archives, describing the process for obtaining a permit to market a species of wild origin. For example, when a bird arrives from his care center, he transfers it to the breeding farm with a CIC of "source W for Wild". This CIC does not give any right to market the bird... for the time being. On the other hand, when two individuals from a W source breed, the offspring will obtain a CIC from an F source, i.e. "captive-born", but not fulfilling criteria C (from a farm "in a controlled environment") or D (from a farm breeding "for commercial purposes"). F is better than W, but still restricts the possibility of exploiting the animal and its offspring. It still needs manipulation, the sublime final touch!

The Grail is when two F-source individuals reproduce. Then, the fruit of this reproduction obtains C-source offspring, i.e. those authorized for trade. It goes without saying that no DREAL officer worthy of the name would ever have validated the issuance of such perfectly illegal CICs. The magic formula we discover sums up the founder's money-laundering strategy:

$$W + W = F$$
$$F + F = C$$

Unfortunately for the clever ones, the DREAL has smelled a snake in the works: unable to trace the origin of the birds, and therefore unable to regularize the tampering it has observed, the birds will be seized. And as is the case whenever a seizure is carried out, it gives rise to politico-administrative-judicial gesticulations aimed either at derailing the operation or, *at the very least,* at deciding on the destination of the animals so that they remain in the bosom of the zoos. At the head of the army in charge of defending the defendants, Sébastien M., again and again, designated as the establishment's capacity holder despite an episodic presence, and the Joan of Arc of lost causes, a veterinarian well known in zoological circles, a certain Florence O.C.

It has to be said that the stakes in this case are much higher than we might have imagined. On the face of it, we're dealing with an illegal establishment, of very modest size and with economic interests of no great importance. But the real stakes lie in the consequences of this seizure. First of all, it will result in the closure of the establishment, and the zoo union can't stand the idea, but above all it will raise the question of the destination of the 70 birds of prey, most of them European, that will be seized. And for the first time in France, the agent at the helm of the proceedings is hoping to comply with regulations by finding a non-commercial destination for these animals, which are banned from sale.

What do you mean, seized birds aren't going to end up in a zoo? Well, no, that's forbidden, and the agent isn't going to seize the proceeds of one offence in order to commit another. And the captivity industry won't like that at all.

In terms of technical resources, the Athénas center was requisitioned for the operation. Lorane and Gilles, these two specialists have rare experience in the field, and are not only able to identify the species present, but also to recognize at a glance whether they are dealing with a male or a female bird with no apparent dimorphism. Above all, they are able to give an age to the animals, which is invaluable in the case of species laundering.

In this case, the entire zoological industry knew that the birds were going to be seized because of their totally botched origin. And, as always, the horde of rabid dogs starts barking to give the affair more prominence, with the unavowed aim of frightening the hierarchy of the environmental police, or even the public prosecutor's office, which is conducting the investigation.

The attacks are *ad personam* of course, a classic, with Gilles and I in the crosshairs, attempting to call into question the technique of capture, restraint and transport. An e-mail discussion was sent to us by the public prosecutor's office, for information purposes, in which we were accused of trying to sell the seized birds of prey. And this discussion took place between the director of the Landes hunters' federation, Sébastien M., and the president of the UFCS. Hunters, an illegitimate official and a president covering up illegal bird sales. Aren't we good? The aim is to discredit the operation at all costs, not only to make people forget the illegality of the birds, but above all to try and get them sent to a zoo rather than a care center.

By sending the raptors to a care center, the zoos run the risk of the birds being released back into the wild, which would demonstrate that the background noise, not to say the myth of "a captive bird can't be returned to the wild" has definitely collapsed. And to prevent this, the *lobby* is going to pull out all the stops. At the time of the second intervention, that of the seizure, I am still banned from working with the environmental police by its management, even though this ban is not based on any legal argument, and above all has no grounds whatsoever other than the offence of dirty mouth. But since it's the Athénas center that has been requisitioned by the public prosecutor, and since I'm a member of the center and, what's more, the world's most capable raptor, it's the Athénas center that, in turn, is requisitioning me, and that's something the unfortunates hadn't foreseen. The interesting thing about a *lobbying* operation is that the more the pressure mounts for those involved, the more reckless the defenders of the zoological industry become, no doubt because of the feeling of permanent immunity that usually protects them.

And so it was that, in the middle of a seizure operation, the agent in charge of the case received a request from his superior for an explanation of my presence on site. The problem was that no one other than Florence O.C. knew anything about it, which showed that the vet had a close relationship with the OFB's police directorate and the person in question. The latter contacts the public prosecutor in charge of the case to point out that my presence as a convicted person endangers the proceedings. Of course, I've never been convicted, but as the lie is distilled informally and on the sly, it's impossible to set the record straight. I'm not the only one to suffer from these manoeuvres: a widely circulated letter from the AFdPZ's lawyer denounced the alleged lack of integrity and probity of the agent in charge of coordinating the operation, triggering a functional protection procedure that his superiors refused to cover up, according to the usual practice. In the world of trafficking, the wildest animals are not always those seen in their enclosures...

In spite of everything, and thanks to the courage of the officer in charge of the case, the birds will be taken to the Jura care center, pending a court decision.

*

Today, we can draw two, if not three, lessons from this case.

Firstly, it illustrates once again that a civil servant fulfilling his mission is almost certain to be subjected to internal and external pressures, frustrations and even sanctions, justifying in passing the immobility of the controlling administration.

Secondly, depending on whether the offenders are powerful or wretched, court judgments will make them white or black: faced with the 300,000 visitors to Cerza, enforcing the law is much more difficult than faced with the 35,000 visitors attracted by the feather park. The 69 birds (including the kestrels we had spotted thanks to the CICs granted by David D.) were seized, sent to the Athénas center, appraised to gauge their impregnation and abilities (flight, hunting, etc.). The last of them were released at the end of 2024 after, for some of them, ten years in captivity. Yes, you read that right, released, and this is a major first in France.

This is perhaps the third lesson to be learned from this affair. It would prove once and for all, if proof were needed, that it is possible to release animals born in captivity and kept over a long period of time, whatever the conditions in which they were kept. To claim that, by its very nature, this project is futile, or even dangerous for the animals, is nothing more than a lie hammered home by the zoo lobby, to the detriment not only of the individuals themselves, but also of biodiversity.

In fact, there isn't a single reintroduction program in the wild that hasn't involved captive breeding. This practice means that an animal born in captivity can return to live in the wild, or learn to live in the wild. Only a jailer whose job it is to show off his prisoners can affirm the contrary.

In France, bearded vultures, griffon vultures and kestrels, among others, have also been reintroduced thanks to captive breeding. We can look at the problem from all angles, but with very rare exceptions, which are to be blamed on their captors, a wild animal, even one born in captivity, even one held for a long time, can be returned to the wild, obviously with the necessary precautions.

It has to be said over and over again, to deconstruct this lie that has become almost a dogma: a captive animal can be released.

This is so true that exotic species, introduced into the wild by accident, carelessness or malice, without any of the necessary precautions for reintroduction, have become so comfortable there that they have come to be known as invasive alien species, or IAS. This invasion of exotic species is in itself proof that captive species, released haphazardly into the wild, can be considered invasive. So, yes, a captive animal can be released.

*

One of the most famous examples of species becoming invasive is the Florida tortoise, also known as the "red-eared slider". Banned from import into the European Union since 1997, it is thought to have originated from animals introduced into waterways by owners wishing to get rid of them. Not a very glorious act, but hardly surprising: pet shops used to sell very pretty "adult turtles" in a plastic bean, with a little island and a mini palm tree in the middle. Except that these tiny mens so-called adult specimens were only three or four days old. But when their owners took good care of them, the turtles grew into beautiful beasts weighing 30 cm and 3 kg, which were also not always practical and potentially dangerous. As a result, many fooled and pitiful buyers abandoned their animals in the first watercourse they came to.

Today, the Florida tortoise reproduces, resists frost and heatwaves, and is renowned for bringing its native competitors to heel. It broadens our demonstration: animals born and bred in captivity are not, in essence, incompatible with life in the wild, even when they are not released in its natural latitudes and with the diet *a priori* made for them.

This case study is no exception. The budgerigar—generally a ring-necked parakeet—has invaded the whole of the Île-de-France region and now seems to be spreading to the south of France. In 2021, the LPO estimated that there were over 8,000 parakeets in the Paris region, three times as many as in 2013. The origin of its development in the wild is controversial. Some say a batch of birds was mistakenly released at a Paris airport. The birds have enjoyed their unexpected freedom to such an extent that, on February 14, 2018, a ministerial decree classifying them as IAS slipped the species *Psittacula krameri* into its Annex I. From now on, "the introduction [of these birds] into the natural environment, whether intentionally, through negligence or carelessness" is prohibited.

And those who maintain that reintroduction is impossible are systematically captive "experts" with a vested interest in ensuring that this dogma is erected as truth. For example, if wild animals are seized, zoo representatives will not only do everything in their power to recover the individuals, but will also assert that this is the only solution, since they cannot be released back into the wild. In this way, they stand a good chance of:

- add some fresh meat to their collections;
- eventualize the episode to create buzz on their social networks;
- monetize it through *merchandising*.

Business is relentless, but that's *business*, and seized animals are no exception to the rule—they're even an integral part of the *business*.

Regularly, the battle resurfaces. Thus, in May 2024, Damian Aspinall, director of the foundation of the same name, renowned for its ability to release wild animals back into the wild, denounced

“3,000 violations of animal treatment rules” on 162 mammal species in “around 30 European zoos” that are members of the European Association of Zoos and Aquariums, including the Boissière du Doré zoo in Loire-Atlantique. And the leader insists on the heart of his emotion: not the existence of zoos, but the fact that those “who run these establishments” and claim to “demonstrate excellence in species conservation” are, in reality, “hurting animals by creating the myth of conservation while demonstrating a great lack of knowledge”.

No, wild animals do not live better in a cage, however large, than in the wild. Yes, a wild animal is destined to live in freedom, even if this freedom brings no profit to the “zoocrats”, as Damian Aspinall puts it, and even if it entails the risks inherent in wild life.

And for those who would question the legitimacy of the Aspinall Foundation’s positions, it has proved that this zoological premise is nonsense. It claims to have massively reintroduced animals born in their zoos, including black rhinos in Tanzania, 70 gorillas—that’s absolutely enormous, seventy gorillas, especially as these specimens have reproduced and, at last count, given birth to 35 additional individuals! -In addition, there are primates in Indonesia, European bison in Romania and Spain, and leopards in Cambodia.

The example of the Parc de la Plume raises the question of the *business of* seized animals, the principle of which I will develop a little further.

18.
The Power of Zoos

In France, animal trafficking is an activity that often takes place quietly, in the territories, but is always thought out, prepared and coordinated on a large scale in the capital. And so it is that in 2018, at a time when biodiversity is finally starting to become a topic of discussion at the National Assembly for no reason or reason at all, a draft decree modifying the principle in place since 2004 is put out to public consultation.

Nicolas Hulot is the French Minister for Ecology, and while he's not the most ecologically-minded among us, he does have the advantage, in my opinion, of having become one as a result of a growing awareness of the issue, and since then he's been very active on the subject.

Hence my astonishment at reading the draft decree, which gives me the impression of living through a nightmare. Up to now, the regulations have been far from perfect, but they had the merit of being in place and only needed to be strengthened, but now... this is one amendment too many.

I take the time to read for several hours, carefully comparing the two texts, the old and the new, and looking for the point at which there's a scam, because there definitely is one. At the helm, as is often the case, is Jacques W., who makes this text even more suspect.

To ensure that the democratic process (which is a big word when you consider the context) is respected, the Ministry carries out a public consultation with the general public, communicates the draft decree to the associations that sit on the CNPN, and publishes it. This is to respect the procedure. In reality, the Ministry, and therefore Jacques W., couldn't care less about the opinions of the public, let alone the CNPN, and *in the end,* the decree will be published and applied. Well aware of the process and well in advance of the CNPN meeting, I published a 20-page document explaining to the mass of sleeping ecologists exactly what this decree is and its consequences. I've included clear tables so that everyone, despite their amateurism on the subject, is fully aware of what's at stake. Any association president with this document in hand would immediately go up to the capital to meet Hulot, which would have been an easy thing to do. Well, nada, nothing, no reaction. Oh yes, sorry, one, the head of legal affairs at the LPO, who, speaking to one of my contacts at FNE, wondered whether this guy, talking about me, was really serious.

To be clear, the text in question opens the door to trafficking in an unparalleled way and rolls out the red carpet for traffickers, so much so that one of the members of the CNPN questioned Jacques W. on the subject, who replied without restraint: "it's my retirement present". And what a present it was! Jacques W.'s sense of impunity is such that he can indulge in this type of outing without any risk and, above all, without any scruples.

Abolition of the possibility of establishing genetic analyses in administrative controls, abolition of part of the obligatory registers, simple remote declaration for holding protected species. In all, 230 species have seen their protection regime reduced, and 60 species, previously subject to the requirement of a certificate of competence, are now free to keep.

The scandal is such that I'm mobilizing everyone I know who might be in a position to lodge an appeal against the decree when it comes out. In the end, it will be part of the future Rewild coalition that will take the text to court.

At the same time, to make sure we don't overlook anything, we're asking a member of parliament to agree to put two questions to the Minister in the National Assembly. In response, Brune Poirson, Secretary of State for Biodiversity, will reassure the assembly by reading out her answer, no doubt prepared by the administration issuing the draft decree. Oh yes, at this point, it should be understood that lying by civil servants in order to pass a lousy text is commonplace, since politicians in their great cowardice will never go and check. Fortunately, I have obtained one of the preparatory documents drawn up by Gérard B., which in itself is a demonstration of the deceptive strategy put in place to force the text through[51]. Perhaps this document will one day be used by to confound the bad guys if a parliamentarian, feeling cheated, were to call for a parliamentary inquiry into the trafficking and the officials covering it up—we can always dream.

In desperation, we passed the message on to Nicolas Hulot who, without meeting us at the Ministry, nevertheless blocked the release of the text.

He resigned shortly afterwards, and on October 18, 2018, the Journal Officiel published the most permissive decree on the subject that France has ever seen. As a result, in France today, it is possible with disconcerting simplicity to help oneself in the wild, falsify markings and launder illegally held European species with little or nothing to stop the traffickers. And all this with the approval of all France's nature protection associations.

51. https://www.sauvage.bzh/sommaire/?file=Chapitre18/null.pdf

As I write these lines, the LPO is launching a "nature in mourning" campaign. You're right, Allain, and nature has been in mourning for a long time now, ever since your organization sold its soul to the devil, hired communicators rather than naturalists, and since you've been in bed with the zoos. It's not for lack of warning, but what credit can we give to an expert who's been smeared by the ministerial administration and the environmental police?

*

Wild animal trafficking in France has two lungs: breeders and zoos. It's the latter that we're now going to focus on. However, to fully understand the issues at stake, we need to be aware of the history of the sector.

Three quarters of the zoos currently on the market emerged between the 1970s and 1990s.

Some establishments achieve a high level of notoriety, whether long-lasting or short-lived. Fallen stars include the Saint-Vrain zoo, founded in 1974 and closed in 1999, and the Val d'Hérault zoo, which closed a year after its creation, even before completing a full fiscal year. Probably the shortest history of a zoo in France. In 2025, Beauval will be the most famous and influential zoo in France. It boasts over a million visitors a year and sales of over a hundred million euros by 2022.

Some zoos are run by local authorities. For example, the Parc de la Tête d'Or, owned by the city of Lyon, boasts "almost two million visitors a year" by 2021, thanks to its free admission, and the Lille Zoo is aiming for 300,000 admissions by 2024, roughly the same number as the Citadelle, home to the Besançon Museum's zoo, another municipal structure. In addition to these municipal facilities, there are the public establishments belonging to the Paris Natural History Museum. The most famous in this category is undoubtedly the infamous Menagerie du Jardin des Plantes, which welcomes almost 500,000 visitors a year, according to figures published by the institution in January 2023; and the oldest remains the Parc zoologique de Paris, better known as the Vincennes zoo, and its 700,000 visitors in 2022, created for the 1931 colonial exhibition and renovated through a public/private partnership (PPP) in 2014.

These establishments are exceptions. As a rule, zoos are private institutions, some of which are headed by third-generation managers. They may be owned by families, such as the Delord family for Beauval (founded in 1980), the de La Panouse family for Thoiry (founded in 1967), or the Gay family for the Bioparc-zoo de Doué-la-Fontaine; but they may also be owned by pension funds, which definitively debunks the myth of the not-for-profit zoo, as it's easy to imagine that if speculators invest in the sector, it's unlikely to be for the beauty of the gesture. Prudentia Capital owns the Amnéville zoological park. Formerly owned by Stéphane Da Cunha, the zoo in La Flèche, Sarthe, was bought in 2017 by Looping Group, whose main shareholder is Mubadala Investment Company, the sovereign wealth fund of the Emirate of Abu Dhabi, also owner since 2015 of the safari park in Port-Saint-Père, near Nantes[52]. Other captivity sites are owned by large corporations, such as Parc Astérix, which until 2022 featured dolphins and sea lions, and which is now owned by a subsidiary of Compagnie des Alpes and its billion euro turnover.[53]

Such characteristics are significant: basically, the zoo is a private enterprise that aims, by definition, to make money. If one accepts the game, this is no disgrace in itself, but, within this framework, it is

52. https://www.lesechos.fr/2017/11/le-zoo-de-la-fleche-repris-par-le-groupe-looping-187696
53. https://www.compagniedesalpes.com/sites/default/files/widgets/reference_document/2025-02/COMPAGNIE_DES_ALPES_DEU_2024_FR_MEL%20%281%29%20%281%29.pdf, p. 4.

obvious that biodiversity, conservation and animal welfare cannot be at the heart of its project and philosophy.

*

Whatever their location or euphemistic name ("Biopark", "safari", "animal park", etc.), the purpose of zoos is to present captive animals to the public. Private zoos have the added ambition of making profits and distributing dividends. Contrary to the narrative trumpeted by the press, today's zoos are never the brainchild of enthusiasts who woke up one morning and said: "Hey! Let's save some little critters!"

Some zoos are notorious for having launched gigantic campaigns to capture felines, sea lions, penguins or monkeys, with appalling mortality rates and serious consequences for the natural environments thus devastated. Moreover, the practice is not always concealed. Some predators even dare to present these massacres as benefits for mankind. In 1968, to justify his purchases of primates "in southern Cameroon", Claude Caillé, founder of the Palmyre zoo, explained how, for him, Africa is essentially a breeding ground for animals, from which it is good to do business, because it supports the local population. In his words:

> "The solution is not to condemn zoos, which are blamed for all the world's ills. One day, affluent nations will have to share their wealth with underdeveloped ones. Safeguarding humanity, whether yellow, black or white, should take precedence over animals. Only then will we be able to talk about protecting nature, because the inhabitants of this country will no longer need to kill wildlife to feed themselves"[54].

54. Claude Caillé, *Mon zoo, ma vie*, éditions du zoo de la Palmyre, 1989, p. 14.

By plundering Africa's natural resources, ripping apart local biodiversity to turn it into cash, zoos would become benefactors of global humanity, regardless of race or continent. Even in the 1960s, you'd have had to have had them tied up pretty tight to dare write, sign and publish such a declaration!

*

The 2000s saw a reversal of fortunes. Zoos wanted to give up their animal-consuming side to meet the new demands of public opinion. The Delord family took control of the only surviving zoo union, and began incredible *lobbying* efforts to have zoos recognized as respectable environmental institutions. Goodbye entertainment; hello conservation!

The way we think about wild animals and treat their captivity obviously remains the same. Only the rhetoric changes, but that doesn't matter. Those responsible for the disappearance of animals in their natural environment now present themselves as benefactors of biodiversity, yesterday's executioners claiming to be today's saviors.

This trick doesn't stand up to a simple examination of the facts. For this book, for example, I compiled data from 86 zoos to sketch out the number of animals held by this deadly industry. According to the official figures I was able to glean, more than 180,000 wild specimens are held by those who try to cover their greed for money with a phony ecological gloss. These 86 establishments attract more than 21 million visitors a year and generate sales of 400 million euros. By 2023, France will have "almost 400 animal parks", with entrance fees set at "between €15 and €30 per person"[55] (by 2025,

55. https://www.francebleu.fr/emissions/l-eco-des-vacances-l-eco-des-loisirs/zoos-et-parcs-animaliers-la-sortie-preferee-des-francais-7678981

adult tickets will be well over €30, with Beauval costing as much as €41). It's not impossible that the actual number of animals kept in captivity by the French zoo industry is somewhere between 200,000 and 220,000.

Like all large, well-structured industries, and therefore well-infiltrated among decision-makers, zoos don't just accept standards: they modify them to their own advantage, precisely to maximize their profits. In the name, for example, of job preservation, ecological awareness and regional development, they often demand and obtain regulatory or tax changes that give them an increasingly free hand. Incidentally, the real culprits are not the entrepreneurs who make the demands, but the politicians and civil servants who give in to their injunctions!

*

After an aggressive communications campaign to promote their virtuous "new mission", it's time to change the restrictive law governing zoos. In 2004, French regulations were made considerably more flexible, and the notion of "conservation" was introduced into the regulatory texts in the name of harmonization with a European directive. Nothing very restrictive in fact, just an obligation to promote the genetic diversity of captive populations.

It was during this period that the influence of zoos, represented by their union, exploded. To understand this, we need only consult the Association Française des Parcs Zoologiques (AFdPZ)'s own documents to see that the organization is no more and no less than a self-assumed *lobby*. Its aim is, on the one hand, "to ensure the representativeness [of French zoological establishments] before the supervisory authorities and international organizations" and, on the other, "to promote them to the media and the general public", "promotional" tools designed in part to ease the conscience of visitors.

The AFdPZ is systematically invited to every symposium, meeting, colloquium or ministerial consultation on the subject of biodiversity, animal welfare or the fight against wildlife trafficking... These participations enable the AFdPZ to flaunt their pseudo-conservation mission and establish their authority in the field, seducing the audience with a rhetorical flourish on conservation that leaves them skeptical, but in which everyone wants to believe, in the name of "it's better than nothing"[56].

The bluff works so well that the French Ministry of Ecology has almost become a zoo salesman. You only have to visit the Ministry's website to understand that zoos are positioned as conservation specialists, not as exploiters of living things.

This is why, when Barbara Pompili announces an end to the use of wild animals in circuses, she does so at the Vincennes zoo. No comment.[57]

This positioning enables zoos to occupy an unduly vested position, where they wield real power of influence to demand and obtain tolerance in many areas in the name of conservation. Ah! "Conservation", the word is out. It is therefore possible, in the name of conservation, to present and sell species protected by CITES, whose trade is prohibited, simply because the Ministry of Ecology, or rather some of its officials, have granted zoos "conservatory breeding" status, for educational and non-commercial purposes[58]. With sales of over 400 million euros and profits of almost 40 million euros, the Ministry of Ecology considers these establishments to be non-commercial. This consideration alone reveals the game being

56. https://www.sauvage.bzh/sommaire/?file=Chapitre18/communiqueMNHN.pdf

57. https://www.liberation.fr/environnement/climat/barbara-pompili-je-connais-les-obstacles-je-savais-a-quoi-mattendre-20210225_SYZFDSFRJVHKPJHMEGC7IOYVXM/

58. https://www.sauvage.bzh/sommaire/?file=Chapitre18/Formation%20AFDPZ%3ACITES.pdf

played by those who roll out the red carpet for *lobbies*, whose true face must one day be seen.

Buoyed by effective communication strategies, the zoological trade operates on a series of carefully constructed lies, like pillars of support for the cause. But these pillars are slowly beginning to crack, simply because some people, myself included, have refused to be lulled to sleep by the flute music played by the president of AFdPZ.

Anything goes to spread these lies. Special programs, episodic series, poster campaigns and even advertising: Carglass[59] offers to repair your windshield during your visit to Beauval Zoo. By dint of bludgeoning the positive message of zoos and the lies to justify their visit, the public, little inclined to reflect on the subject, allows itself to be seduced and fooled.

59. https://actus.zoobeauval.com/carglass-repare/

19.
The Illusion of Conservation

Of all the zoo industry's lies, conservation is undoubtedly the biggest one, the one that wins it the most believers, brings it the most benefits, and gilds a papier-mâché blazon with fine gold. In this respect, I think the message has got through to the public: everyone is convinced that a zoo is Noah's Ark, without which animals would have disappeared.

In theory, conservation is a safeguarding operation, the result of which benefits the species or the areas they inhabit.

To qualify as such, species conservation means preserving genetic diversity and behavioral aptitudes. Genetic diversity is preserved in particular by the number of individuals held, ensuring a cross-fertilization that avoids inbreeding and genetic drift. The preservation of natural behavior is ensured by keeping the animals in areas close to their original environment, in zones where the tranquillity of the residents can be respected. Here, animals are confronted with rivalries and dynamics of group domination, territoriality, foraging, predation, sexual competition... The absence of these criteria makes zoo conservation a grand illusion.

Zoos are incapable of meeting the essential conditions demanded by conservation, both from a genetic point of view and in terms of the territories made available to the animals, whose ridiculous size leads most of them to develop behavioral pathologies that single-handedly jeopardize conservation operations. These pathologies are so widespread that to name this "captivity-related madness" we speak of stereotypy, which is so much prettier. And to try to overcome this madness, strategies called "enrichment" are implemented, managed by ethologists. Ethologists and enrichment are the two buzzwords in zoo semantics to convince people of their professionalism and show how much they care about their animals' well-being. But, whatever words we might use to dress it up, this madness is a fact, directly induced by keeping wild animals in captivity, and a very significant number are affected by it.

In the zoo, each birth is the subject of a well-oiled communication campaign, and regularly, for the occasion of an artificial insemination, the birthday of a resident of an "exceptional" age or the arrival of a new animal, a layer is added, sometimes to the point of indecency, as on the occasion of the arrival of pandas or golden monkeys at Beauval zoo. The message is clear: zoos save species by enabling their reproduction. Not a summer goes by without a zoo special (always the same ones) extolling the virtues of this disinterested conservatory.

What zoos call conservation is in fact a European breeding plan (EEP) for the management of individuals in captivity. Let's make no mistake: EEPs are programs designed to prevent a father from mating with his daughter, or a mother with her son, by setting up a *studbook* and appointing a coordinator. It's a collection management plan, as has been practiced for 200 years by national stud farms for horses, without anyone ever calling it conservation.

The many species, for which we are extolled for breeding, are generally very common in the wild. In a study published in 2012, the Born Free association already highlighted that only 17% of the species present in zoos were endangered species[60]. The young animals bred in these establishments are frequently euthanized due to excess[61] or poor genetic origins, as was the case at Copenhagen Zoo in 2014. A giraffe and four lions were euthanized on the pretext of genetic uninterest after, of course, the establishment had profited handsomely from the success generated by the presence of these young animals.[62]

Even EAZA, the European zoo union, admits that between 3,000 and 5,000 perfectly healthy animals are euthanized every year in Europe for these reasons[63]. Quite apart from the ideological aspect of this management method, which contributes to the big lie about the conservation role of zoos, this management is totally illegal, particularly on national territory. In France, article R 655-1 of the rural code stipulates that: “The fact, whether publicly or not, of deliberately killing a domesticated or tamed animal, or one held in captivity, is punishable by a fine of the 5th class”[64] (from 1,500 to 3,000 euros in the event of a repeat offence), the life of a wild animal is not worth much.

If you delve into the *stud books* of species, it’s easy to see that the genetic quality of the individuals reproduced is of such mediocrity that, in many cases, they cannot be returned to the wild. Indian rhinoceros, koalas, red pandas, Persian panthers, okapi, Asiatic lions and many others have a level of inbreeding that jeopardizes

60. https://www.sauvage.bzh/sommaire/?file=Chapitre19/BELGIUM.pdf
61. https://www.ouest-france.fr/leditiondusoir/2018-01-15/le-zoo-fait-...e-lionceaux-il-en-euthanasie-61d3d320-d9ef-4bce-9231-76abfb1e6323
62. https://www.sauvage.bzh/sommaire/?file=Chapitre19/LemondeGirafon.pdf
63. https://www.20minutes.fr/planete/1334250-20140326-girafon-quatre-lions-pourquoi-zoo-copenhague-euthanasie-animaux; https://www.rtbf.be/info/societe/detail_3-a-5-000-animaux-euthanasies-chaque-annee-dans-les-zoos-europeens?id=8211000
64. https://www.legifrance.gouv.fr/codes/article_lc/LEGIARTI000006419579

their chances of survival in the wild, and creates a risk of species degeneration. This rate of inbreeding partly explains the low viability of certain individuals and the high mortality of young.

A few examples? The okapi, a giraffe from Central Africa, is totally symptomatic of the gap between rhetoric and reality. Ten okapis are kept in four establishments in France. All ten have the same ancestry. Under these conditions, it's hard to continue to argue in favor of conservation in order to justify showing them to the public. This species is classified as critically endangered, and its wild population is estimated at less than 50,000 individuals. At Beauval zoo, a leading establishment in this field, okapis are presented in treeless grassy enclosures, despite the species' forest-like behavior. Individuals sadly wait for their daily ration outside the lodge door from 4 p.m., a fine example of behavioral conservation.

The situation is undoubtedly even worse for Indian rhinos, 18 of which are kept in six zoos in France. All the animals are the offspring of a couple from Bassel Zoo in Germany. In Europe's other zoos, 75 individuals are held in 28 different zoos. Only seven of these 75 individuals are not the offspring of the German couple. The Indian rhinoceros is classified as vulnerable by the IUCN, and is only represented by 2,575 individuals in the wild.

Even the once-common king of animals is not immune to this phenomenon. A study published in the journal *Mammalian Biology*[65] shows that for the Asiatic lion, "the core population of the program includes individuals from sibling or half-sibling unions, including all founding females". Between 2007 and 2009, almost 70% of the 57 cubs born in this program died within four months, the majority after just a few weeks. The cause was inbreeding between the offspring of the nine "founder" lions.

65. https://www.sauvage.bzh/sommaire/?file=Chapitre19/MammalianBiology.pdf

This genetic poverty is at the same time a godsend, since it justifies capture operations in the name of safeguarding the genetic purity of the species. Incidentally, these operations involve using the animals by presenting them to the public, which is totally incompatible with conservation actions.

One thing is certain: the genetic quality of captive livestock in Europe will not allow the return to the wild of species held.[66]

In addition to inbreeding, which can eventually spread over several generations (the famous "new blood"), interbreeding and hybridization normally prohibit any return to nature. Taking subspecies into account is a fairly recent development, and up until the 1980s, gorillas, giraffes, chimpanzees, lions, tigers and parrots were commonly mixed between subspecies.

Such is the extent of interbreeding and inbreeding that nowadays one of the "big jobs" of these establishments is to study genetic origins in order to consider removing the "impure" ones for conservation. Unfortunately, biodiversity won't have time to wait for this "sorting" to be carried out, and some species will soon only be found in captivity, reinforcing the position of zoos as last-chance establishments.

Beyond the smokescreen of conservation through breeding, we must not confuse birth with reproduction. Reproduction takes into account the entire life cycle, from birth to the young animal's ability to reproduce. For reproduction to be successful, the number of births must exceed the number of deaths, which is rarely the case. A birth is simply the delivery of a youngster, not the most difficult stage in a species' life cycle.

As important as births are (but on so many species with no conservation issues, are they really), the complete reproductive cycle is much less obvious. Take, for example, symbolic species with

66. https://www.lexpress.fr/informations/le-faux-alibi-des-zoos_624822.html

real conservation issues, such as the African elephant, and you'll see that the rate of viable births is extremely low, not least because of the conditions in which they are kept. In this species in particular, premature mortality is considerable. Of the 223 African elephants that died in European zoos, over 35% did not reach sexual maturity at the age of 15. 72% of them were captured in the wild, and only 30 reproduced, producing 89 young, of which only 34 are still alive.

Then, of course, there are the reintroductions, the zoos' new hobbyhorse and communication strategy. Father David's deer, the Arabian oryx, the Hawaiian goose, the European bison, the griffon vulture, the bald ibis and Przewalski's horse are the seven emblems of the reintroduction *success story* achieved thanks to zoos.

This success story needs to be tempered, however, firstly because in most cases, the animals were bred in specialized breeding establishments using individuals recovered from zoos. We can therefore speak of participation in reintroduction, but little more. And secondly, because for several species, the genetic quality of the reintroduced species is totally questionable. For example, the reintroduced population of Przewalski's horses is made up of thirteen individuals. By the way, they had become extinct due to high demand from zoos. The inbreeding of this species is, among other things, the cause of significant morphological changes, making the long-term success of this reintroduction uncertain. This species could even become a symbol of the failure of the zoo model of *ex situ* conservation.

At present, the pressure on zoos to strengthen their conservation alibis is so great that the biggest ones are going straight for a return to the wild, with plenty of press for fans to say, "Ah, you see... they're doing conservation!" Two gorillas from Beauval Zoo, for example, have returned to their homeland[67]. This return must also be put into

67. https://actus.zoobeauval.com/article/premiere-reintroduction-de-gorilles/

perspective, and compared with the reality of the figures. Firstly, they were initially released in a three-and-a-half-hectare sanctuary, not "in the wild"; and secondly, this "program" concerns only two gorillas out of the 17 held at Beauval. But above all, these are two animals released out of 35,000 presented to the public at this zoo. The Aspinall Foundation, which initiated the program, has already reintroduced 60 gorillas. The return of two gorillas from Beauval is only a tiny contribution, which would require *at least* a little more humility.

Be that as it may, and in the rare cases where zoos do implement conservation actions, they are by no means heroic operations, but merely compensation for the damage perpetuated over almost 100 years of capture and hunting to enrich collections for often private profit. In my opinion, this participation is a fair return on investment and deserves neither glory nor a medal.

20.
The Cash Cow

In addition to the EEPs that zoos put forward, they also argue in favor of funding *in situ* conservation programs.

Although conservation funding is one of the zoos' main arguments, it is difficult to quantify, and the amount of the contributions is often obscure. In fact, in view of the profits made by some of these establishments, particularly the larger ones, the funding they earmark for conservation amounts to crumbs thrown to associations, like food for Parisian pigeons. WAZA (the World Association of Zoos and Aquariums), which brings together some 1,300 institutions, declares that it funds conservation programs to the tune of 350 million francs (or just under 79 million euros) every year. When applied to WAZA's 1,300 member institutions, this amounts to just over 60,000 euros per zoo, or less than 0.50% of their annual sales[68]. This sales calculation does not take into account *merchandising* and catering, which account for almost 30% of the budget. EAZA (the European Association of Zoos and Aquariums), for its part, claims to have financed five million euros

68. 700 million visitors with an average ticket price of €25 = €17,500,000,000 = 0.45% of sales.

over 15 years for some one hundred participating zoos, or €3,333 per zoo per year.[69]

By way of example, the details of conservation operations and their financing, presented in the reports of the Beauval Nature association, speak volumes about the fund-raising and communication techniques used to sell the conservation benefits of zoos to the gullible.

Why Beauval? Probably because it's the most talked-about, the biggest, the best, the only, in their own words, and it's the biggest budget allocated to conservation for France, at least on the surface. A reading of the Beauval Nature association's 2017, 2018, and 2019 balance sheets reveals, first, that Beauval Zoo has never donated more than €176,000 to conservation, through its association. That was in 2019[70], a year in which the establishment posted a profit of €4,467,000.[71]

In 2017 and 2018, not a single cent was donated to conservation[72] by the zoo, despite a profit of €3,580,600 for 2017 and €4,538,200 in 2018[73]. A closer look at the association's budget reveals that it is made up entirely of donations. These donations come either from a few zoos or from private partners. The zoos contributed €28,600/ year over the three years under review, and the private partners an average of almost €410,000/year. AAB, Aeb, API Surveillance, Aux Charpentiers de France, CMF and many others contributed over 50% of the budget. The latter, whether named patrons or donors,

69. https://www.parcanimalierlabarben.com/animal/campagnes-eaza?cn-reloaded=1
70. https://beauval-nature.s3.eu-central-1.amazonaws.com/pdf/rapport_activites_beauval_nature_2019.pdf
71. https://www.sauvage.bzh/sommaire/?file=Chapitre20/BeauvalComptes2019.pdf
72. https://www.sauvage.bzh/sommaire/?dir=Chapitre20/Rapport2017-18
73. https://www.sauvage.bzh/sommaire/?file=Chapitre20/Beneficebeauval2017-2018.pdf

are often among the zoo's suppliers. So, is the notion of disinterestedness required for tax exemption respected?

It also states that a sum of almost €110,000, donated in kind by Beauval, represents the work of employees involved in the Beauval nature association, the famous "valorized volunteer work". Although the technique is not necessarily illegal, it does raise the question of the motivation of the staff who, for the time being, invest in the association during their working hours, calling into question the notion of voluntary association work, and the motivation of the zoo, which charges the association for the hours worked on conservation—there's no such thing as a small profit.

One fact sums up better than any other the extent to which conservation is a pretext and not a real conviction: Beauval has created a "conservation house". Until 2021, this was in fact a 4 m^2 wooden kiosk for a 44-hectare zoo run for leisure.[74]

And if I had to rub it in, I'd add that Beauval is also four hotels with a total of 1,400 beds, representing an investment of 42 million euros and an advertising budget of 2.5 million euros per year[75]. By way of comparison, it costs 250 euros to buy and save one hectare of equatorial forest[76]. The Beauval dome, a new attraction in 2020, would have financed the acquisition of 160,000 hectares of this treasure trove of biodiversity[77]. Conservation is above all a question of priorities...

These monstrous investments, required to keep animals in captivity, undermine the argument that the cost of reintroduction justifies so few of them, compared with the cost of maintaining them in captivity, which is costing Beauval Zoo €125,000 per day for 2019.

74. https://www.sauvage.bzh/sommaire/?file=Chapitre20/Maisonbeauvalnature.png
75. https://www.sauvage.bzh/sommaire/?dir=Chapitre20/Beauvalentreprise
76. https://www.lemonde.fr/planete/article/2007/06/28/un-site-pour-acheter-et-sauver-les-forets-tropicales_929241_3244.html
77. https://www.lesechos.fr/pme-regions/centre-val-de-loire/le-zoo-de-beauval-investit-40-millions-dans-un-dome-tropical-1169986

21.
The Aura of the Panda

Zoos don't just lie about their contribution to conservation. They also lie when they claim they don't buy animals. They barely admit to trading...

The AFdPZ (French association of zoological parks), whose primary mission is—let's face it, it's easy to guess—conservation, has taken the position that "no sale"[78] is not for sale! It's in the name of this virtual gratuity that the French Ministry of Ecology has granted zoos a safe-conduct allowing the issuance of a specific CIC, through which the use of protected species without sale is made possible[79]. So, let's put the church back in the middle of the village. Firstly, by definition, zoos sell animals, insofar as they sell tickets (and more) to their visitors. Two, contrary to what they claim, zoos actually rent, buy and sell animals, which is not at all criminal if the source is legal. Three, since these establishments rent, buy and sell animals, how can they claim to contribute to conservation?

Between 2000 and 2005, eight French zoos bought 14 white rhinos from Africa, representing a market worth over 320,000 euros.

78. Publication AFdPZ website.

79. https://www.sauvage.bzh/sommaire/?file=Chapitre18/Formation%20 AFDPZ%3ACITES.pdf

Europe is not to be outdone: 55 individuals of the same species were traded in 33 European zoos for a total of almost 2,000,000 euros.

Between 2008 and 2019, Beauval zoo alone either bought or sold more than 25 different species. A giant anteater, three tigers, two lions, a jaguar, a Sichuan takin, two white rhinoceroses, a California sea lion, but also a Moluccan cockatoo, three Philippine cockatoos, three Pesquet parrots, a snake-eater, and so many others... All worth several hundred thousand euros. These sales were made either between zoos, or from Beauval Zoo to private individuals and vice versa.[80]

Of these dozens of species, many are purchased in the animals' country of origin. Some are even bought in pet shops, such as a spectacled caiman from the Ferme Tropicale in Paris. In other cases, sales are made through suppliers, some of whom have already been convicted of offences involving the sale of protected species.

It would take too long to draw up an exhaustive list of the number of animals traded by French zoos, but one thing is certain: the trade argument no longer holds water.

They are not the only ones to sell animals, and some do so openly. Loro Parque in Tenerife, for example, through its foundation, reproduces rare parrots and sells them under the pretext of financing their protection. It's a bit like opening a brothel to get prostitutes off the street, but it doesn't make much sense.[81]

JACO ANIMAUX, a totally illegal establishment in the north of France (no authorization to open and no certificate of capacity), supplies French zoos. With him, a sloth will cost you €4,250, a collared anteater €5,000, a pair of fishing cats €4,750... He's not the only animal supplier in France. International Zoo Services, Zoological Services, IWELS, Zoo Ekzo are just some of the suppliers

80. https://www.sauvage.bzh/sommaire/?dir=Chapitre21/Zimsvente
81. https://avesdojorge.com/wp-content/uploads/2021/03/LPF-28-05-2025.pdf

who offer several thousand animals every year, from canaries to African elephants, chimpanzees, orangutans, tigers, lions, giraffes...

This is so common, especially for birds, which have virtually become a consumable commodity, with mortality rates approaching 100% for some species, that EAZA, through its Zim's platform, provides a list of 400 importers, traders and pet shops for its member zoos.[82]

This list even includes the names of people who have sometimes been convicted on several occasions or are wanted by Interpol[83]. The proximity between traders, traffickers and zoos leaves no room for doubt as to the role these establishments play in the trade, and by extension trafficking.

Beneath the lie of non-trade lies the lie of lies. With the economic expansion of emerging countries, we are witnessing an explosion in the construction of zoos, in the image of the excessiveness of those built in China, a country that has been criticized for the way in which animals are kept there. It's perhaps one of the worst places in the world to be a captive animal. Animals thrown alive to predators, starved, escaped—not a year goes by without Chinese zoos being splashed with scandal.[84]

Yet China is currently embarking on an unprecedented operation to purchase animals, particularly from Africa. These animals could potentially end up in European zoos through exchanges in the name of conservation. In this way, they enrich their "collection"

82. https://www.sauvage.bzh/sommaire/?file=Chapitre21/importateurEAZA.pdf

83. https://www.sauvage.bzh/sommaire/?file=Chapitre21/Iwels.pdf ; https://www.leparisien.fr/seine-et-marne-77/un-trafic-d-animaux-exotiques-demantele-31-03-2004-2004873979.php

84. https://www.huffingtonpost.fr/international/video/en-chine-coursepoursuite-avec-un-le-opard-echappe-d-un-zoo_181266.html ; https://www.linfo.re/monde/asie/719909-chine-insatisfaits-les-investisseurs-d-un-zoolancent-d-autres-animaux-aux-tigres

with wild animals taken from the wild, despite the warnings and recommendations of the IUCN.[85]

Elephants, once again, are a real stake for these zoos, which take advantage of a pretext of overpopulation in certain countries such as Zimbabwe, are sold for a life in captivity again in the name of conservation. Between 2012 and 2019, 67 elephant calves left Zimbabwe for China, despite a CITES resolution opposing the trade. The contract is worth several million dollars, and scientific resistance and specialist protests have not succeeded in stopping these departures, given the economic stakes involved.[86]

But emerging countries aren't the only ones importing elephants to keep them in zoos. The elephants at Beauval Zoo come directly from the wild via an English zoo, which acquired them in Africa, while at the same time advocating the preservation of this species—what a contradiction.[87]

Added to this is the fact that the commercialization of wildlife is not limited to the purchase of animals. Animals can also be rented. The most famous and most political are the pandas, which Beauval officially rents out for €800,000 a year[88]. Their arrival in France has enabled the zoo to increase its visitor numbers from 600,000 to 1 million, and the zoo continues to grow, with support from the entire political class, and I was going to say almost all categories. To keep its giant pandas for another 10 years, a zoo in Taiwan paid 11 million

85. https://www.sauvage.bzh/sommaire/?file=Chapitre21/CourrierIUCN.pdf
86. https://www.leparisien.fr/societe/le-triste-commerce-des-elephanteaux-vendus-a-la-chine-30-10-2019-8183160.php https://www.aa.com.tr/fr/afrique/le-zimbabwe-vent-des-éléphants-à-la-chine-malgré-les-protestations-des-écologistes/502046
87. https://www.zoobeauval.com/zooparc/liste-des-animaux/elephant-dafrique
88. https://www.bfmtv.com/economie/les-pandas-de-beauval-machine-a-cash-ou-gouffre-financier_AV-202108020186.html

euros[89], and the Pairi Daiza zoo will have to pay a rent of 1 million dollars a year to keep its own, but it's worth it.[90]

Despite all this, no one ever questions these establishments, in the name of a rather mysterious deification. Even when you use technical, scientific and factual arguments, there is always a category of believers who will defend these practices whatever the cost.

---

89. *Ibid.*
90. https://www.bfmtv.com/animaux/zoos-le-business-des-pandas_AN-201402240098.html

22.
The Elephant's Head

The third lie promoted by zoos concerns the myth of the longevity of animals in captivity. Yes, in captivity, animals live in peace. No hunters, no predators, just the good life! Basically, all wild animals should have the chance to be imprisoned...

In 2020, to prove just how unbearably untrue this zoo cliché is, I used Zim's database to list all the African elephants (over 300) and Asian elephants (over 400), over 200 white rhinoceroses and over 400 lynxes held in captivity in France over the last 50 years. I calculated the average age they were when certain specimens died, and compared it to the average life expectancy of their species in the wild.

The result is that, on average, African elephants live 27 years in captivity, compared with 56 years in the wild. Asian elephants live 18 years in captivity and 41 years in the wild. White rhinos survive six and a half years in captivity—their natural lifespan is fifty years. Lynx live four and a half years behind bars, compared with eight years without fences. These figures, compiled from the zoos' own written declarations, are indisputably violent.[91]

91. https://www.sauvage.bzh/sommaire/?file=Chapitre22/longeviterzoo.pdf

As long ago as December 12, 2008, a study published in *Science* came to much the same conclusion[92]. In it, Canadian researchers compared "the life expectancy of elephants in European zoos with those in Kenya's Amboseli National Park and those used in Burma's logging industry. The results show that elephants kept in the wild or working in the wild have at least twice the life expectancy of their cousins.

The study shows that African elephants live 16.9 years in captivity, compared with 56 years in the wild. Even taking into account poaching-related mortality, the longevity of free-living animals remains almost twice that of their captive counterparts, at 35.9 years. The same is true of Asian elephants, whose life expectancy in captivity is 18.9 years, compared with 41.7 years including pack animals.

And yet, "the life expectancy of captive elephants has increased"... The main author of the report concludes: "Protecting elephants in Africa and Asia is much more effective than protecting them in Western zoos".

Of course, it could be argued that the two ages are not comparable: the first is a statistical average, the second is an estimate of potential, which does not take into account the reality on the ground (if poachers are rampant, the average age of death in the wild falls, while the estimated lifespan remains high). But this objection is based on the utmost bad faith. Life expectancy in the wild is a physiological reality that can be scientifically verified. Life expectancy behind bars is a statistical reality. In both cases, it's a question of estimating how long an animal can live under the conditions to which it is exposed; and the result, based on the declarations of the zoos themselves, is so blatant that it should lead to the immediate closure of establishments presenting animals to the public.

92. https://www.lapresse.ca/sciences/en-vrac/200812/12/01-809939-les-elephants-vivent-plus-vieux-dans-la-nature-que-dans-les-zoos.php

In other words: dear zoos, when you take elephants from the wild, you are not contributing to the conservation of a species, you are not saving animals, you are shortening their life expectancy by half. You're not helping to conserve the world's genetic heritage, you're destroying it. It's one thing to entertain spectators. If laws and regulations are respected, no one can take offense at the fact that you're selling tickets, *fun*, *goodies*, popcorn and overnight stays. Everyone is free to be a customer or not. On the other hand, if you lie to the public, that's something else, and that "something else" is not acceptable.

Because it's not just pachyderms or felines. All wild animals suffer and die in captivity. Orcas, for example: no, these animals don't live longer in aquariums than in the wild. At the end of 2016, we learned of the death of Granny, an orca over a century old[93]. In captivity, the life expectancy of these animals doesn't exceed fifty years, and generally peaks at thirty-five years...

*

Faced with this statistical and scientific evidence, you'd think that zoos would abandon the misleading argument of longevity and summon their communicators to fabricate another, more biased but less questionable one. To do so would be to underestimate them. To support the false myth of longevity, the captivity *lobby* has a trick up its sleeve: it relies on isolated cases such as the elephant at Vincennes zoo, which stands out as an exception and enables the idea of long life in captivity to continue to be propagated.

His name was Siam, and he was the largest Asian elephant held in a zoo, spending the last 33 years of his life at Vincennes. Victim

93. https://www.sciencesetavenir.fr/animaux/animaux-marins/la-plus-vieille-orque-du-monde-en-liberte-est-morte_109468

of an injury to his front leg, directly induced by being kept on an inappropriate substrate, he was euthanized at the age of 52 on the pretext that "his head was too heavy for his legs", declared the veterinarian in charge of his euthanasia.[94]

This statement illustrates the pride of animal keepers who are incapable of questioning themselves by recognizing the detestable conditions in which they are kept, and who will always manage to turn a failure into a success, by pleading their exceptional longevity and large offspring (14 young). But once again, you only need to zoom in to see the reality of the situation.

The Vincennes zoo has held eight other elephants, all of which died prematurely, between the ages of 0 and 38. And of Siam's 14 descendants, only three are still alive today. Siam is now enthroned in the grand galerie de l'évolution at the Muséum national d'Histoire naturelle in Paris, stuffed and accessible on payment of an entrance fee of €10[95]. Even dead, it will still bring in a little money, which will be used to finance the abysmal 247 million euro abyss that the Vincennes zoo will cost taxpayers, a cost generated by an aberrant restoration that is dangerously draining the budget of Buffon's noble zoological institution.

The reality is that, just as one swallow doesn't make a spring, one individual among hundreds of others doesn't make a statistic. With a few exceptions, in captivity, even when supposedly "well-treated", the longevity of a wild animal is less than half its life expectancy in the wild. This is an indisputable fact, with figures to back it up.

*

94. https://www.letemps.ch/societe/exstar-cirque-knie-siam-contemple-desormais-paris-oeil-cristal
95. https://www.sauvage.bzh/sommaire/?dir=Chapitre22/Siam

So, to counter the figures, more and more zoos are boasting about the expert resources they deploy to try and "help endangered species". In fact, it's all a question of semantics, because what zoos pass off as a success (births by artificial insemination) is in fact the result of a detention policy that is an abject failure. When a species, such as the rhinoceros or the elephant, fails to reproduce naturally, it's proof of a dysfunction largely linked to captivity. Human intervention to artificially inseminate an elephant may represent a veterinary feat, but it is by no means a guarantee of success.

In fact, the scale of the failure is commensurate with the investment involved. On an international level, a cryobank for elephants has been set up to prevent the disappearance of this species from American zoos. Specialists estimate that such an extinction could occur within 40 years, if nothing is done to curb inbreeding.[96]

In some cases, assisted reproduction has disastrous consequences, as in the case of this black rhinoceros who died at the age of 28 (a young male) in 2018 at Doué-la-Fontaine zoo. For the purposes of artificial insemination, he was anesthetized with a view to collecting his sperm by electroejaculation[97]. Killing a species in an attempt to reproduce it is more failure than success. "Fortunately, like Siam the elephant, Kata-Kata the rhinoceros will end up stuffed to strengthen the collection of the Bordeaux Natural History Museum.[98]

This zoo has already been in the news on several occasions, notably following the deaths of two lionesses killed by a male

96. https://www.lepoint.fr/sciences-nature/les-nouvelles-arches-de-noe-16-08-2012-1696525_1924.php
97. https://www.sauvage.bzh/sommaire/?dir=Chapitre22/Katakata
98. https://actu.fr/nouvelle-aquitaine/bordeaux_33063/bordeaux-kata-ka... oir-est-arrive-aujourdhui-museum-dhistoire-naturelle_29461302.html

during contact, and three cheetahs following inappropriate deworming.[99]

On the other hand, very few establishments—and these are the exceptions—provide their residents with enclosures of an appropriate size, to the point where it becomes difficult to stop reproduction.

Not only the size of the facilities, but also the cohabitation with other congeners or, on the contrary, the isolation of males who are only brought into contact with each other in the heat of the year, help to approximate the conditions and lifestyle of the animal in the wild, thus encouraging reproduction, particularly in the case of large species. At Cabarceno, south of Bilbao in Spain, for example, a facility created in a former quarry covering almost 700 hectares provides comfortable spaces for animals such as rhinoceroses, hippopotamuses and elephants. They are housed in enclosures up to twice the size of Beauval Zoo.

This zoo has the world's largest herd of captive African elephants, numbering 18 individuals. The herd comprises six males and 12 females. Of these 18 elephants, 13 were born at the zoo. This zoo is a mirror image of the cramped conditions of most French zoos and their failure to reproduce, which they try to hide behind the smokescreen of artificial insemination and sperm banks.

99. https://actu.fr/pays-de-la-loire/doue-en-anjou_49125/bioparc-doue-_-deux-lionnes-quelques-jours-apres-leur-arrivee-zoo_30305085.html https://france3-regions.franceinfo.fr/pays-de-la-loire/maine-et-loire/maine-loire-deux-femelles-guepards-sont-mortes-au-zoo-doue-fontaine-mai-dernier-1846462.html

23.
An Unnatural Marriage

The omnipotence of the captive breeding industry is largely due to the almost total absence of contradicts and counter-power. In theory, this role should be played by environmental associations, which are, however, with rare exceptions, totally absent from the debate. Much more serious than this absence, however, is the almost historical proximity between the world of nature protection and the world of nature ownership, which has led some of the sector's behemoths to sell their souls to the devil. Not out of Machiavellianism, no, simply out of financial interest. As the associations don't understand the traffic, the stakes and the players, they don't see zoos as a danger, so why not collaborate?

Moreover, this collaboration, not to say this dubious mixing of genres, is not new. You only have to go back a little in time to realize that keeping wildlife and protecting it are intimately linked, and the subject is often treated in concert by the specialized press of the time. Reading magazines such as *La vie des bêtes* highlights this phenomenon. At the same time, the magazine dealt with endangered animals (already) and Jean Richard and his lions, despite the fact that they were kept and exploited in a circus. There's even an unbelievable classified ads section featuring monkeys, parrots, lions and tigers for sale.

This mix of genres almost served as the basis for the construction of the French associative world, some of whose pioneers are still active today. Looking back at the genesis of the LPO (Ligue pour la Protection des Oiseaux—League for the Protection of Birds), we understand that this mix of genres is almost an institution. One of the first presidents of the LPO was Jean Delacour, who served as president from 1910 to 1968[100]. He was, however, one of the first creators of a private zoo, in 1920, when he opened the Parc de Clères, which housed over 3,000 animals, all captured, imported and presented to the public.

He supplied the natural history museums of Paris, London and New York with thousands of stuffed animals brought back from his many colonial expeditions. In 1969, Jean Delacour became the first president of the Association Nationale des Parcs et Jardins Zoologiques privés (ANPJZ), the forerunner of today's French zoo union.

This mix does not call into question the many written works and the knowledge of living things they have enabled, it simply explains that animal protection and zoos are historically linked, without anyone today questioning the coherence of this link.

This link will endure for a long time to come, since Jean Delacour's successor at the head of the LPO is none other than Antoine Reille, founder of France Nature Environnement, vice-president of the CNPN (Conseil national de la protection de la nature) and director of the WWF. Antoine Reille also worked with François de La Grange on the program Les animaux du monde, which he produced after the latter's death, alongside his wife Marlyse de La Grange. François de La Grange also collaborated with Konrad Lorenz in the creation of Branféré Zoo.

---

100. https://www.sauvage.bzh/sommaire/?dir=Chapitre23/Delacours

At a time when animal protection goes hand in hand with animal husbandry, it's hardly surprising that this mix of genres is rooted in the collective unconscious. And yet, all the zoos for which these pioneers of nature protection collaborated were skilfully supplied with captured animals, of which only 5% arrived alive in France. These same imports were occasionally denounced by the naturalists in question.

The problem is that this mixture of genres continues to this day, unchallenged. Today, the LPO and the Beauval Zoo are delighted that the latter has become one of the largest LPO refuges in France[101], after the Doué Zoo, the Barben Zoo, the Paugre Safari, the Sables-d'Olonne Zoo[102] and many others to come. However, the LPO's official position on[103] zoos sets out a list of conditions for collaboration, including: "Any breach of existing regulations (quality of infrastructures and conditions for the keeping and presentation of animals, absence of any removal of animals from the wild, etc.) would render any prospect of future partnership unthinkable, and any current partnership null and void".

The collaboration between the LPO and these zoos flouts these principles and is in total contradiction with the latter's ethics, proof of the zoos' great power of persuasion, no doubt due to their shared history.

When you read between the lines, you realize that the Beauval zoo is committed to funding conservation programs run by the LPO, which raises the question of the disinterestedness of such a decision.

101. https://actus.zoobeauval.com/article/beauval-refuge-lpo/
102. https://www.sauvage.bzh/sommaire/?dir=Chapitre23/LPOZOO
103. https://www.sauvage.bzh/sommaire/?file=Chapitre23/position-LPOZOO.pdf

The current director of the Île Grande zoological station, the LPO's flagship care center, is also in touch with the world of captivity, as a former director of the Lille zoo, and is involved with EAZA, the union of European zoos.[104]

Under these conditions, it becomes impossible to mobilize the LPO, FNE and others in actions against the deregulation that encourages trafficking, insofar as, even when explained in a didactic way, the consequences of the relaxation of the law are not perceived as a danger, since zoos are allies rather than enemies.

The phenomenon is such that Allain Bougrain Dubourg, current president of the LPO, was vice-president of the Vincennes Zoo Renovation Committee until its reopening in 2014. He is still an advisor to Océanopolis, and a director of the Muséum national d'Histoire naturelle[105], owner of the menagerie at the Jardin des Plantes, whose existence and conditions of detention are in total contradiction with regulations and the principle of animal welfare.

The LPO, which is one of the partners of the pompous Académie de Fauconnerie du Puy du Fou and the ANFA (Association Nationale des Fauconniers et Autoursiers de France) of raptor trainers, whose activities should in principle keep bird protection associations at a distance. ANFA's partners also include the UFCS and, worse still, the OFB, which is supposed to monitor.[106]

It's hardly surprising, then, that nature conservation groups don't take offence at the commercial use of animals, some of which even come from their own care centers, since the pretext of their conservation will be used as an implacable argument and, for

104. https://www.sauvage.bzh/sommaire/?file=Chapitre23/CVRMLPO.pdf
105. https://www.sauvage.bzh/sommaire/?file=Chapitre23/PortraitABD.pdf
106. https://www.fauconnerie-francaise.com/nos-actions-de-protection/ https://www.sauvage.bzh/sommaire/?file=Chapitre23/IMG3196AFPDF.jpg

zoos, collaboration with these associations protects them from any recourse.

It was in the name of this same mix of genres that, in 2004, the Nicolas Hulot Foundation opened its School for Nature at Branféré Zoo. Learning how to protect the environment while offering a visit to an animal prison... it's clear that mixing genres and conflicts of interest have a bright future ahead of them.

The combination of fraud and fraudulent techniques, the lack of resources and knowledge on the part of the control authorities, the corruption of certain officials and the inability of associations to act as a counterweight all contribute to a flourishing trade that everyone seems to be happy with.

24.
Seized Animals: Captivity Industry's Heist

Taking care of trafficked animals is a priority if we are to put an end to part of the wildlife trade and attempt to offset its impact through rewilding operations. This subject has guided my entire career, and today I know almost everyone involved. I saw the birth of almost all the structures which, under the pretext of receiving seized animals, created zoos without the animal's interest ever having been the priority in their initiative. Above all, I was the first, back in the 1990s, to put the problem of seized animals squarely on the table.

The vagueness maintained by the Minister on the subject led to the creation of more or less shaky establishments, some of which closed down for lack of resources. Most of these establishments are still illegal, have opened illegally, or are operating illegally in terms of national or international regulations, without any administrative intervention. The impossibility of the State taking charge of seized animals, especially in large numbers, protects these establishments from administrative or judicial intervention. All it takes is for a celebrity to be named honorary president, and the little courage left in the administration disappears at the mere mention of his name.

In my thirty-five-year career in this sector, I've long understood that the handling of seized animals is a lucrative business, the preserve of a few, which allows two or three animal protection foundations to make a living from the media, the players in the field to be satisfied administratively, and zoos to be enriched in the name of "animal welfare and protection". However, under French law, a seized animal is said to be "in the hands of the law". Article 706-145 of the French Code of Criminal Procedure "prohibits any use" other than the search for the truth during an investigation. An animal in the hands of the law remains the property of the offender, who is therefore responsible for its upkeep and preservation. The perpetrator bears the financial burden, even when the seized animals are entrusted to a third party.

De facto, any sale or exhibition to the public, under any pretext whatsoever, of a seized animal is therefore prohibited, and constitutes illegal use and misappropriation of evidence punishable, under the same Code of Criminal Procedure, by three years' imprisonment and a fine of 375,000 euros.

It is only once a court decision has been handed down that confiscation can take place, transferring ownership of the "thing seized" to the State. Article 131-21-1 specifies that confiscated animals are to be entrusted to a foundation or association for the protection of animals recognized as being of public utility or declared. In the spirit of the text, therefore, there can be no commercial exploitation of confiscated animals, and yet.

In France, for many years now, everything possible has been done to ensure that seized animals are almost systematically sent to zoos, disregarding the Code of Criminal Procedure and the regulatory context, and jeopardizing every procedure should the offending parties lodge an appeal for annulment. And the most scandalous aspect of this commercial organization created around seized animals is that some of the players are animal protection foundations or associations, several of which have even made a "specialty" of managing this category of residents.

Obsessed by the desire to intervene in the event of a seizure, but unable to do so without zoo cages, these foundations are prepared to make any concessions, happily ignoring the consequences of their actions, thereby fuelling the flourishing "seized animals" *business*. This *business* is based on an unnatural union which leads these foundations to finance zoos when they set up "special facilities for seized animals", all of which will end up behind the bars of their rescuer. In this way, they co-finance, among other things, establishments that are just as questionable as each other, at least in terms of the final destination reserved for the animals concerned.

A veritable schizophrenic organization has been set up. The reception of seized animals is organized by animal protection foundations which rely exclusively on those who exploit, buy and sell them, in total violation of French and European legislation. For CITES (applicable in Europe through the European regulation) is very clear on the subject: only species born in captivity from legal parents can be traded. These species benefit from what is known as source C, which authorizes their trade. No other source may be traded.

It is this same regulation that lists commercial activities, including the presentation of animals to the public. Contrary to what some French officials at the Ministry of Ecology claim, CITES considers that the presentation of animals to the public is indeed a commercial activity that must comply with the same rules as any other.[107]

A confiscated animal or its offspring should therefore never be able to benefit from authorizations allowing its commercial exploitation.

However, the French authorities have granted a safe-conduct to zoos, ensuring that they agree to take charge of seized animals in

107. https://www.sauvage.bzh/sommaire/?file=Chapitre24/RCEE33897.pdf

return for the right to exploit them commercially. This sauf-conduit allows controls to take place, justifying to CITES that France is fulfilling its obligations in this area, and at the same time making zoos the exclusive recipients, since they alone have the right to commercially exploit animals banned from trade. It's vicious, but terribly effective.

Against this backdrop, the situation regarding seized animals is one of incredible chaos, with no centralized information system, and where animals are entrusted to establishments not on the basis of their skills, adaptation or philosophy, but on the basis of local arrangements between operators and certain control authorities. Add to this the unjustly acquired notoriety of certain foundations positioned as intermediaries in the capacity of specialists, and it's all over. The animal is never treated as a treasure of biodiversity to be restored to its original state, but as a marketable commodity that must end up, whatever the cost, in the aviaries of private breeders or large zoological collections.

This phenomenon is exacerbated by the complexity of a seizure operation, which sometimes prompts control officers to ignore the issue and entrust the animals to foundations with little regard for the animals' future, but which at least provide the "complete" service.

This *business* of seized animals gives rise to battles of interests orchestrated for years by the ministerial administration, refusing to create specific structures that leave room for anarchy and do-it-yourself, in the face of which zoos pass for professional rescuers. This chaos also protects the zoos from any claims by associations against the Ministry, as the players in the field are too busy protecting their territories and personal interests, and waging war against each other for "the cause".

Year after year, with the closure of the international market and the difficulty of sourcing animals, the reception of seized animals has become an important issue, enabling zoos to bring back species that are sometimes very rare into their collections and to complete them by replacing the dead, at little or no cost.

The recent bankruptcy of the Val d'Hérault zoo confirmed the phenomenon in broad daylight, with the 30 Millions d'Amis Foundation admitting, through its representative on camera, that the rarest animals will obviously go to the Beauval zoo.[108]

When the news of its liquidation was circulated, I immediately contacted Arnold L., the "specialist" from the Fondation 30 Millions d'Amis, to whom I sent a long e-mail to explain to him the extent to which in this zoo, there are species whose rarity merits a return to their country of origin *at the very least*, and in any case that these birds should be treated in a manner commensurate with their conservation status.[109]

By way of reply, Arnold L. tells me that he "dealt with Beauval and that it's simpler that way". The birds were therefore sent to the zoo, which took advantage of the opportunity to enrich its collection with animals that had lost all hope of ever returning to their original homeland, and some of which had obviously already died. This operation will also give the Beauval zoo the opportunity to orchestrate a fine communication campaign and pass itself off as an animal rescuer.[110]

As the foundations are not the managers of the facilities, they are not in a position to guarantee the future of the animals they have been entrusted with, but they can easily and without constraint take charge of them, without having to do what is the hardest thing of all—ensure their upkeep and day-to-day management.

108. https://www.dailymotion.com/video/x3k0y96

109. https://www.sauvage.bzh/sommaire/?file=Chapitre24/Mail30million.pdf

110. https://www.lanouvellerepublique.fr/loir-et-cher/commune/saint-aignan/deux-cent-vingt-oiseaux-sauves-au-zoo-de-beauval

No one really knows why these foundations, whose initial speciality is domestic animals, are entrusted with such rare animals when not a single one of their members holds a certificate of competence, they don't manage any structure, and their competence, given the numerous documents and videos available on the net regarding the handling and restraint of wild animals, leaves a lot to be desired.

Communication has such incredible power that it has succeeded in making French courts think that these foundations are experts in the management of seized wild animals. The Fondation 30 Millions d'Amis, in particular, has become AFdPZ's almost exclusive transporter, supplying it with animals for commercial exploitation whenever the opportunity arises.

In France, four operators share the "cake" of seized animals. Le Refuge de l'Arche, Tonga Terre d'Accueil, La Tanière-zoo refuge and Elephant Haven.

All these operators are directly and intrinsically linked to the captivity industry, even if they call themselves "refuges".

The refuge de l'Arche, located in the Mayenne region of France, is typical of these excesses. Christian Huchedé, its founder, was an employee of the commune of Château-Gontier, whose good faith is beyond doubt. At first, he was entrusted with wild animals from the European fauna, then he was asked to take in small primates and felines. Little by little, his "refuge" began to take shape. His mission was to take in and care for "animals seized by the authorities", whom he considered, on principle and in his own words, to be "too impregnated with man and unable to be released back into the wild". As a result, the refuge has become a zoo with "almost a thousand boarders of one hundred and twenty different species" for around 15 euros in 2025. In return, the animals that enter die, regardless of their conservation status.

Tonga Terre d'Accueil, for its part, does not make presentations to the public, making the establishment more respectful of regulations. However, it infiltrates almost all of its residents into zoos, which raises the question of legality and ethics, as it is directly subordinate to the Saint-Martin-la-Plaine zoo, which is the investigator. Sporadically, a few individuals are sent to sanctuaries in South Africa. Decisions to sanctuarize or keep in captivity this seized or abandoned fauna are rarely submitted to real specialists in reintroduction, but, as is often the case, entrusted to those who have an interest in keeping the animals in captivity.

Elephant Haven is a sanctuary in the Limousin region of France. It aims to be a retirement home for elephants. Created on the dogma that "a captive animal cannot return to the wild", the animals housed there will be condemned to end their days in captivity and presented to the public, at least in the initial version of the project.

The latest addition to the quartet, La Tanière-zoo refuge is, like the refuge de l'Arche, a zoo that presents seized or abandoned animals to the public, and thus follows the same commercial logic. It was created solely for this purpose, playing on the fact that the public is more inclined to visit a refuge than a zoo. Managed by an SAS, La Tanière was set up by a businessman who is surfing the seized or abandoned animal *business*.

An authorization to open was issued in January 2018. Yet La Tanière suspiciously obtained a derogation a year earlier to keep bears and sea lions in conditions that did not comply with their biological requirements... where the sea lions were kept in an above-ground garden pool[111], and the bears regularly walked around on a leash.[112]

111. https://www.sauvage.bzh/sommaire/?dir=Chapitre24/Otarie
112. https://www.sauvage.bzh/sommaire/?file=Chapitre24/Ourslaisse.jpeg

Between champagne watering for the bears and television for the monkeys, the circus-style management of the animals allows for the development of aggressive communication via a television series on C8, the images of which should make any wildlife specialist jump.

La Tanière will be directly impacted by the Covid crisis, which highlights the establishment's commercial dependence, and will lose, barely a year after opening, its "capacitaire", who, rumor has it, has not been paid for many months and prefers to withdraw. The loss of the manager made it illegal to operate the establishment. However, *lobbying* by the founder, and the OFB services that support him, resulted in a visit from the Minister of Ecology, Barbara Pompili[113], accompanied by over 300,000 euros in subsidies[114]. To bring things full circle, Sébastien M. has become La Tanière's new manager.

Almost all these establishments are co-financed by the Fondation Bardot, the Fondation 30 Millions d'Amis and One Voice. All of them are dedicated to the fight against animal exploitation, commodification and trafficking. Could it be that their existential needs on the media scene have largely superseded their originally laudable aims and philosophies?

The commercial exploitation of seized animals is all the more outrageous given that the usually convoluted regulations are, for once, crystal clear:

"A refuge or sanctuary for captive wild animals is a non-profit establishment that takes in animals of non-domestic species that are or have been captive, have been seized or confiscated, found abandoned or placed voluntarily by their owner who wishes to relinquish them.

113. https://www.ecologie.gouv.fr/barbara-pompili-et-marc-fesneau-visiteront-zoo-refuge-taniere-en-eure-et-loir-vendredi-5-mars-2021

114. https://www.bfmtv.com/societe/ferme-a-cause-du-covid-19-le-zoo-r...-se-bat-pour-sa-survie-et-appelle-aux-dons_AN-202103160276.html

Any activity involving the sale, purchase, rental or reproduction of animals is prohibited.

The presentation of animals in the form of shows and any direct contact between the public and the animals at the initiative of the visitor or the staff of the refuge or sanctuary are also prohibited".

A sanctuary that shows animals to the public is no longer a sanctuary, it's a zoo. Back to square one and the previous chapters!

*

The crux of the problem is not even ethical or ecological: it's legal. The law is, or should be, the only effective bulwark against trafficking. And beyond the legal context, there are the conservation aspects that should push all players (authorities, associations, shelters, countries of origin) to take a disinterested stance on the future of seized animals. Depending on their origin, their genetics and their rarity, they obviously do not have the same heritage value and, under no circumstances, can they be managed like domestic animals.

This is why, when a wild animal is collected, the first thing to define is its status. Its fate will depend on it. For example, if an authority seizes a kangaroo, a Korean squirrel or a ring-necked parakeet, its patrimonial value is much less important than that of a tiger, a macaw rubrogenys or a gibbon. These animals deserve to be integrated into a real captive breeding program and then reintroduced into the wild.

But, as the doctrine preached by captive industry professionals is: "a seized animal cannot be returned to the wild", the question obviously doesn't even arise. What angers me most about this popular rumour is that it's not based on any objective, technical or scientific element, it's just a rumour distilled constantly and to whoever wants to hear it. So much so that anyone who says the opposite, with arguments, evidence and a little experience, is systematically made to look like an idiot.

*

Each of my interventions in the field of seized animal management, some of which received considerable media coverage, ended up putting the spotlight on the subject to such an extent that today nobody can deny that there is a real placement problem. Well, we're going to deal with it, but in our own way. In a new farce, the Ministry, taking advantage of the "Revision of the EU Action Plan against Wildlife Trafficking", has created SAASS, Services d'assistance aux animaux sauvages saisis, and guess who is in charge of managing this service? The AFdPZ. Do I really need to comment?[115] The purpose of this service, as presented in the brochure, "is in line with objective 14 of the European Union's action plan (revised in 2022) against wildlife trafficking, which aims to improve access to care for seized or confiscated specimens". The brochure in question is nothing but an apology for captivity and breeding, and says nothing about return to the wild, even though the famous Objective 14 is very clear on the subject, calling for "intensifying efforts, where appropriate, to effectively reintroduce seized live specimens into the wild".[116]

The reality is that the behemoth of animal presentation to the public is now in a position to redistribute seized specimens among its colleagues, according to the accointances of each. Entrusting a

115. https://actus.zoobeauval.com/afdpz-missionnee-lutte-contre-trafic-especes-sauvages/#:~:text=Créée%20en%201969%2C%20l'AFdPZ,'éthique%20de%20l'Association. https://www.sauvage.bzh/sommaire/?file=Chapitre24/Présentation_SAASS_19.03.25.pdf

116. https://www.sauvage.bzh/sommaire/?file=Chapitre24/PlandactionCEE.pdf

zoo with the management of animals seized by the courts is like being concerned about the condition of cattle and deciding, as Anne Sylvestre sings in *La vache engagée*, to convene a congress on the condition of cows and entrust the presidency to the butcher.

This time, and if anyone still had any doubts, the proof is in the pudding that the Ministry's agents have just sold off wildlife in general and seized wildlife in particular to the captivity industry, without a single association, foundation or even politician lifting a finger. The State's submission to Beauval's diktat is a scandal. Thirty years of fighting to protect wildlife independently of zoos have been more than challenged by this unnatural association, which bears witness to the complicity between *lobbyists* and senior environmental officials. This victory for the zoos is obviously a defeat for the animals, but also for those who, like me, have labored to denounce guilty practices and build independent alternatives that have encouraged traffickers to organize themselves to recuperate the influx of seized animals, from which they derive a substantial profit. Here again, we must not give in to discouragement: anger must fuel our ability to take positive action to stop the massacre!

25.
Rewild: A Major World First

Fighting on, especially when all seems lost, is an attitude I seem to have always clung to. Not because I'm heroic, no, it's worse: because I'm stubborn. Let's not kid ourselves, I've had my down moments, my head in the bucket, my morale in the doldrums. I've been discouraged; I've felt the poison of what's-her-name flowing through my veins; I've had the grey thoughts of someone who feels like they're tilting at windmills without the madness of Don Quixote to help them. Nevertheless, I believe these moments of *blues* have always fuelled my desire to return to the arena. Again, I'm not saying this with a puffed-up chest, or to paint myself as Superman. I'd be more like the cat who comes back the next morning when, every night, you think he's dead! But I'm too aware of the importance of wildlife, biodiversity and the balance of ecosystems to give myself the right to give up and never try again.

So, in 2018, when it's obvious that the Biome is not going to happen, that the Landes hunters have bungled seven years' work, when one of the biggest conservation operations of the last twenty years collapses just as we were reaching our goal, I persist. In this case, I'm continuing my work as an expert, but on the sly, since I seem to have become *persona non grata*. Every time I contribute to having animals seized, it's heartbreaking, because, in the end, I'm working voluntarily to enrich zoo collections. That's when I came up with the idea of getting several associations involved in the cause together to discuss the fate of seized animals. So I organized a meeting in Bordeaux, on the premises of Darwin Ecosystem, with, among others:

- One Voice;
- Sea Shepherd;
- Hisa and its not-so-obvious acronym (Human Initiative to Save Animals);
- the Le Biome association, of course;
- the Athénas Center;
- Wildlife Angel;
- and Darwin Ecosystem, our host.

Over the course of five hours on the day, I summarize what I've learned about wildlife trafficking; I point out the limitations of the various ways of combating zoo management of seized animals; and I propose the creation of a group to set up and manage a center specialized in this field. Each would have its own specificity: investigations, interventions, seizures, management of animals and their return to the wild, communication, organization of events and symposia, mobilization, etc. Six months later, another meeting, this time in Paris. We finalize the creation of a group of associations to be known as Rewild, with the aim of opening a shelter for seized animals. According to our estimates, the project should take two to five years to come to fruition, and require a fund-raising effort of between seven and ten million euros.

The figures may seem colossal for Biome, which I represent, or the Athénas center. On the other hand, for some of our partners who have a lot of money and know how to raise public awareness, there's nothing monstrous about them, especially as the project is likely to strike a chord with Internet users, and therefore generate donations, provided we create content to explain the urgency and necessity of the project. I'm working on writing the articles of association for our coalition and giving the concept some teeth. It was against this backdrop that, a few months later, a rumor reached my ears: the Pont-Scorff zoo in Morbihan was up for sale. I know the place well, having worked there on numerous occasions as an expert and for training courses. So I contacted the zoo's former owner, the founder's daughter, who still owns the land but is no longer in charge of the SARL. She gave me the contact details of Sauveur F., the current owner. The idea is doubly tasty: for environmentalists to buy a zoo and transform it into a sanctuary aimed at rewilding the animals entrusted to its care! Starting from an existing structure rather than from scratch would save us an enormous amount of time and money. I put the idea to the collective. They were enthusiastic.

As a result, I contacted Sauveur F., the owner, to begin negotiations. He's a very special person. A renowned child psychiatrist and court-appointed expert in his own right, he is also the head of a small family-owned group of healthcare establishments. Although he assures me that the rumor is unfounded and that his zoo in Morbihan is not for sale, he agrees to receive me at his château in Saint-Genest-de-Contest, in the Tarn region, because the story I've outlined intrigues him.

*

When I arrived at the château, four hundred kilometers from home, I felt as if I'd discovered a ghostly estate! The lopsided wrought-iron gate is almost bare. In the garden, kangaroos and emus roam amidst car wrecks. In addition to Pont-Scorff, Sauveur F. owns the 3 Vallées zoo in Montredon-Labessonnié. Although he claims to "not particularly like" these establishments, he nonetheless sees them as "the best way to intervene immediately in the conservation of species".

Intrigued by the château, I ask him to tell me the story of the place, which is about as sordid as the décor. Formerly a home for children in difficulty, it was the scene of violence and acts of paedophilia that led to its closure. It was against this backdrop that Sauveur (the name is evocative) bought the site.

He assumes he is incompetent in the technical management of his parks, which he says he became owner of almost by accident. Intellectually, he immediately strikes me as honest and level-headed.

Our conversation lasts a good seven hours. After I present my project, he shows me his zoo, an infamous hellhole built under the leadership of Sébastien M. (him again). The facilities testify, if proof were needed, to the total ignorance of both regulations and the basic needs of animals on the part of the so-called expert we had occasion to mention.

The visit is an opportunity for him to tell me about his history with zoos, and to cross-check the information we both have on Sébastien M., his former employee, and David D., the DREAL Occitanie agent we've already mentioned. It was at this point that Sauveur passed on to me all the documents showing how David D. had framed Sébastien M. to improperly obtain his certificate of competence—it's a small world indeed.

The visit is punctuated by a presentation of the Rewild project, interspersed with Sauveur's anecdotes about his zoo. He is captivated by what I tell him, and seduced by the idea of transforming a detention center into a reception center for seized animals. During the dinner he invites me to, he asks me:

—How are you going to finance the acquisition?

—With a fund-raising campaign on the Internet.

—Can you believe it?

—If I didn't, I wouldn't be here with you.

He thinks, pauses, stands up, sits down and says: "You're right, Scorff [as he called Pont-Scorff] is for you!

He likes the project, he even likes the fact that this ambition is driven by a coalition of impecunious associations, but in the *starting-blocks* to launch a fundraising campaign via the Internet. I think he thinks it's crazy too, and I like to think that's what really won him over. So we parted on an agreement in principle, which was then refined and formalized during meetings in Paris with lawyers.

On December 18, we meet in Lorient to sign the purchase agreement. Lorane du center Athénas and Perrine de Hisa, both recently appointed co-presidents of the Rewild coalition, officially sign the purchase of SCI Fi-zoo, itself a 70% owner of SARL Bretagne zoo. The signing took place after three hours of reading deeds and, of course, the payment of 200,000 euros advanced by Sea Shepherd. This sum is in fact an advance on the repayment of the investments Sauveur F. has injected into the company, because in fact the zoo, incredible as it is, is being bought for the symbolic euro.

It's 9 p.m. when we leave the notary's office, under a very Breton drizzle. This time, it's done, we own the Pont-Scorff zoo, we've fucking done it!

26.
Panther Priority

Ahead of the signing, we presented the zoo takeover project to Hugo Clément, whom we met at a lynx release organized by the Athénas center in the Jura.

While we were having lunch, I explained to Hugo that we should start a fund-raising campaign. He's well versed in the subject, since a few weeks earlier he had asked his audience to finance the transformation of a hunting reserve into a reserve. The result: around 300,000 euros raised. The amount raised seemed exceptional... until I explained to him that our project was going to be a huge success and that, as a result, we should aim for a million euros. To say he was skeptical would be an understatement: "500,000 would be great!"

He's an expert in social networking and communications, but I'm a project man, and having already raised sums in excess of 3 million euros, I feel, I'm sure, I know that it will work. However, I'm also aware of my limitations. Clearly, social networks, digital communication and *crowdfunding* codes aren't my thing. I have my little ideas, somctimes, but I suspect that others have much more experience in the field than I do. So we're going to launch a 500,000 euro fund.

We agree to launch our operation just before Christmas. We talk to Hugo's production team and GoFundMe, the platform we've chosen for our crowdfunding campaign. A shoot is discreetly organized in Pont-Scorff to capture the images needed for the launch video entrusted to the Hugo team. The video is sent to me three days before the publication date, which means that deadlines are tight. When I saw it, I jumped at the chance. The video is 100% anti-zoo and misses the point. The tone, the images, the message are all wrong. Not a word about the project, about the need to rewild, just a long lament about the misery of captive animals and above all a call to buy a zoo to free its residents.

Being anti-zoo doesn't shock me, as we all know; except that, in this case, we're tiny tots up against ogres. We have to be strategic. If we put out a clip that blows the lid off the captivity industry, we'll probably make a lot of money, all right. But it'll be a Pyrrhic victory, because in the medium term, we'll end up with all the establishments on our backs, and they have long arms. And above all, that's not what the project is about.

Our goal is to explain that we're going to buy a zoo and turn it into a shelter, with the aim of rewilding instead of life in the slammer. Our coalition is not intended to attack the animal holding industry as such. If it indirectly enables the millions of park and safari guests to become aware of what underlies this exploitation in terms of biodiversity, animal welfare and, why not, human ethics, so much the better, but let's not confuse the issue...

So here I am, embroiled in some rather bizarre debates where the roles are reversed: Hugo tries to explain to me that I should go for the fight and hit him hard; and me, who's the type to go head-on, urges him to change his approach and go easy on the trigger. I have my say both as the instigator of the adventure and as a professional who has worked hard in wildlife-related activities. So, while I'm no stranger to the art of creating buzz on Facebook, and I don't mind specialists taking the *lead* on these subjects, I do know the zoo *lobby* and the captivity unions and their power to cause trouble. And I believe that, for such an important project, it's pointless, even dangerous, to create considerable adversaries, especially when their activity is not at the heart of our approach.

I'll lose in the great game of democracy. My warnings flop. I don't get my way, and the film is thrown around on all the social networks. The fundraising campaign is a huge success, and the takeover of the Pont-Scorff zoo by the Rewild coalition is a bombshell that sends the captivity industry into a panic, right up to the European Zoo Association (EAZA).

We take physical possession of the zoo on December 19, 2019. At the time of signing, I was appointed manager. As Rewild acquired the SCI that owns SARL Bretagne Zoo, I am the only one with an official role in the company. As such, I am responsible for:

- personnel;
- animal welfare;
- compliance with regulations;
- safety;
- and I must take care to avoid accidents and pollution, which can quickly turn into tragedy in a zoo.

In other words, I'm the captain of the ship. It's up to me to take responsibility for the rogue wave that will soon be breaking over us, not least because of the project's launch advert.

*

Although passionate about the animal world, I've always been convinced of the importance of human relations. Not "human resources", but relationships! Good, because I have to explain to the staff what they're going to be eaten with. I present the Rewild project to the employees, then invite those who want to follow us to stay with us... and the others to make themselves known. This isn't a pressure tactic, it's a necessity. What's ahead of us will require maximum investment from everyone. If anyone feels uncomfortable with the ideal that guides us, it's best to stop right there.

In the days that follow, I sense that the atmosphere is becoming electric in the good sense of the word. On the Net, the fund-raising campaign is in full swing. The caretakers are following this project, on which they will be directly dependent, with great interest. All the employees confide in me that they are exhausted from carrying this ageing zoo in a noxious atmosphere. Legal proceedings, which are legion, also affect their morale. Some of the caretakers have been working here for fifteen years and expect a lot from us. We'll do what we can, but we'll need money to carry out our ambitious project.

In three days, the €500,000 has been raised thanks to many small donations. Marc Simoncini, the businessman who made his fortune with Meetic, announces a donation of €200,000. The media coup was a success. The kitty now boasts €730,000 from over thirty countries. The "Let's buy a zoo" operation is a success, especially as we already own the zoo. Those involved in animal captivity realized this too late. They try to mobilize to make a counter-proposal that would put the nail in the coffin of these hysterical ecologists, but there's no time! But once Hugo announces that the goal (his own) has been reached, donations become rarer.

Too bad, there's no turning back now. On the contrary, we need to move on to the active phase of the project. Without delay, the question of closing the zoo to the public is put on the table. Obsessed with the success of the "kitty" operation, we had not broached this essential subject. Perrine and I decided against immediate closure. Weird? More realistic. Between food for the animals, salaries for the caretakers and outstanding bills, Pont-Scorff is expensive to run. I propose that we opt for an educational approach: the public would be invited not to visit a park, but to see animals to help us pay for their return ticket. In this way, the meaning of our approach would be more concrete and more accessible, and we'd be able to recover a few not-useless picaillons to pay for our running costs.

Gilles and Lorane M., from the Athénas center, and Lamya E., head of Sea Shepherd France, are of the opposite opinion. They think the impact would be greater if we announced the immediate closure of the facility. To reassure us, Lamya repeats over and over that Sea Shepherd will advance the money if necessary, so there's no need to worry about that. All we need to do is communicate well—and they know how to do that. Once again, I'm outvoted. I accept it, because it's the basis of associative democracy. Inwardly, however, I'm divided. I'm as attracted by the principle of immediate closure as I am worried about the financial consequences, even in the short term.

Hugo Clément is sending us a filmmaker who will spend weeks documenting the project, the people and the animals, with a view to making a report one day. Miguel is his name, or rather the nickname I gave him, as I couldn't remember people's first names, and I couldn't remember those of the caretakers either, so I gave each of them a nickname, which for the most part stuck, and strangely enough, I could remember.

This was followed by a myriad of interview requests from all over France and thousands of messages on social networks, some of which set the world on fire. An Internet user hiding under a pseudonym launched a diatribe entitled "Lettre ouverte à Hugo Clément et Sea Shepherd" (Open letter to Hugo Clément and Sea Shepherd), which was immediately picked up by *Ouest-France*. The comments are so defamatory and reflect such rare incompetence that I don't think the pseudonym in question even imagined that his rag would be such as to create a buzz. After investigation by the caretakers, it turns out that the author is an educationalist by the name of Raphaël L., in charge of conserving a collection of... dead animals at the Bordeaux Natural History Museum. These remarks are aimed directly at me, calling into question my probity, my honesty and my skills. Only my hairstyle was not the subject of a comment, and with good reason. The tone is set, the serious business begins.

*

I set up my office in an old meeting room. I put in sofas, hammocks, tables and chairs for sharing meals. I want to invent a convivial place that is both my office and the staff's space. Not to control, on the contrary: to physically mark the need to be together in this same challenge. I'm delighted to see the staff make the place their own in the blink of an eye, aware that what's at stake at "Scorff" is beyond us, and must therefore transcend us. The animal captivity industry is so furious that the EAZA *newsletter*, published a few weeks after our announcement, is working to "solve the Rewild problem". This time, the war is well and truly declared.

Enthusiastic or excremental, social networks are on fire, that's their function (if I've understood correctly). I disconnect from GAFAM to concentrate on the mission without letting myself be polluted by certain nauseating comments.

The morning after our arrival, the administration puts its first banderole. The labor inspectorate came to look for lice in our tonsure. The next morning, the DDPP arrived. The veterinary services have a field day. The next morning, the environmental police arrived. The message of this harassment is clear: the zoos have let the wild animals loose, and are hoping to cut us to shreds by unleashing the wrath of the authorities. This confirms their influence over the administration.

At the head of this army, the director of veterinary services, Jean-Michel C., is personally taking charge of the Pont-Scorff case. No, not personally: very personally. Before we arrived, the zoo was already under formal notice for mistreatment. The case concerned two Sri Lankan panthers accepted by the previous management, even though there was no facility capable of housing them. The animals have been confined for months in a squalid, dirty and untidy animal pen. The size of their lodge is less than 15 m^2 per individual. To clean, one animal has to pass into the other's dressing room, stressing both animals and carers.

In just a few days, the dressing rooms adjacent to the animals' were cleared of all the garbage that had accumulated over the years and made available to the felines, so that they no longer came into contact with each other during cleaning operations. They now have two boxes each. A special hyper-secure enclosure has been built outside, and the animals will soon be transferred there. Despite these improvements, the director of the DDPP will continue the procedure for mistreatment, arguing that since it is the SARL that is targeted, having taken its head, there is no reason for the procedure to stop. And less than 15 days after our arrival, we are hit with a sanction for mistreatment, just to further sully our initiative.

*

Not a week goes by without a press article about the zoo, focusing on:

- its disastrous management;
- its mistreated animals;
- or those unconscious *vegans* who've taken it into their heads to swindle gogos on the Internet.

This was the moment that the OFB chose to put an end to the adventure of the Gabonese greys. It was one of those many mornings when I was walking to the zoo with my head in my shoulders, wondering what was going to happen to us today, that the receptionist warned me of the presence of the environmental police at the entrance to the zoo. No sooner had I gone down to meet the officers than I was told that they had come as part of a procedure to return the grey parrots to the WPT.

—A restitution procedure, I say? You mean a seizure?
—No," he replies, "a restitution procedure.

I'd heard a lot of garbage about this in my career in general, and since our arrival in particular, but this time I have to say they managed to surprise me. A few days earlier, I had gone to the Landes to collect the birds and put them in a quarantine room before allocating them a large aviary to await their release, because they were ready to leave. I had overlooked the energy our opponents would put into preventing us from achieving this feat, which would definitively establish the credibility of our initiative. The famous restitution procedure is in fact an attempt to convince me, via a police department, to let the birds leave for England, and to do so, the ministry has set up a detour of procedure. As you can well imagine, I'm no fool, and I'll sweep this bogus procedure aside out of hand, since a police force, even an environmental one, can never intervene in a private disagreement, especially without a court decision.

No matter. Hours of explaining, attesting and proving that I'm indeed responsible for these birds, and that I'm in charge of looking after them until they're back in the wild, will do nothing. At most, I managed to gain time, as the environmental police agreed to suspend their farce until the next morning, when I was summoned to the gendarmerie for a hearing. They will have had one night to try and invent an offence that will allow them to seize the birds. And that offence would be: illegal transport of a protected species. Unlucky for them, I know the regulations on the subject perfectly well, and birds that aren't intended for commercial use don't need any authorization, since only transport for commercial purposes is regulated.

Faced with bad faith, abuse of power and scheming, it's hard to fight back. I decide not to play the rebel by blocking the doors, preventing the entry of the truck that has come to pick up the birds, hoping that all will go well for the future. If only I'd known. In retrospect, and given the way we were treated afterwards, I regret it. So the birds were rushed off by the WPT, who shipped them to England. I understand that these birds will never see their homeland again, and that's probably what saddens me most.

The prosecutor decides not to prosecute me, but offers me a reminder of the law, which I refuse. I want a trial, I want this scandalous abuse of process to come to light, which is obviously not to the liking of the corrupt agents. You can imagine the newspaper articles and the comments on social networks. A procedure will be launched to have the seizure annulled, and it will be won, but what's the point, the birds have flown.

*

Against this backdrop, and despite the long working hours, I set about drawing up an inventory of the situation, particularly as regards the animals, so as to ensure that no illegalities committed prior to our arrival could be attributed to us.

The results were indisputable: the establishment had a combination of non-compliant facilities, non-compliant conditions of detention and safety, and non-compliant identification for most of the animals. 45% of the mammals inspected had no provenance, no papers or no markings. It's even worse for birds. 82% have non-compliant or anarchic markings, making it impossible to prove their legal provenance. And all this has been kept quiet by the DDPP for years. However, the establishment, which has undergone four inspections in 2018, three in 2019 and three in February 2020 alone, making it the most inspected zoo in France. It makes you wonder whether the director of the DDPP has lost all common sense and, in any case, all notion of equal treatment of fellow citizens, an essential concern for a senior civil servant. We'll soon have the answer.

No sooner had the audit report been submitted to the authorities and forwarded to the public prosecutor, than we received an order to close the site, with a scarcely believable list of recitals. Ill-treatment of animals, operation of an establishment without a certificate of capacity, increased mortality, pollution of the waters of the Scorff, I could go on and on. The copy of this decree is sent to all the e-mail inboxes of all the ÀPQR editorial offices and local TV stations. In the words of a France 3 journalist, he had never seen such a desire to harm and such a lack of neutrality on the part of a departmental director. Well, that makes me feel better, I'm not paranoid then.

In less time than it takes to write it, we're being dragged through the mud and accused of a thousand horrors, for which we're going to spend long hours on social networks trying to re-establish a semblance of truth and reassure the few partners we have left.

This could have gone on for a long time had it not been for the intervention of the zoo's vet, a chap close to retirement age who, at first, showed a real distrust of us, but now finds our initiative rather sympathetic. In any case, he recognizes all the efforts made and the energy spent on behalf of the animals. The unfounded harassment from the veterinary services department finally got to him too. Without being asked, he provided us with a sworn statement to the effect that our arrival had been accompanied by a drop in animal mortality, an improvement in their living conditions and an adjustment of the feed to their specific needs.

In short, a statement diametrically opposed to that of the hysterical senior official. It goes without saying that we were quick to pass on this attestation to the press, which, for once, reported it widely.

Immediately, a minor miracle occurred. We were invited to meet the prefect in person. Far from trembling, I rejoiced at this prospect, which I'd come to stop believing in. From the very first day of our arrival, and even before the first checks, I had asked for a meeting with him, but he had never deigned to reply.

*

On the day we met, the local edition of *Ouest-France* ran the headline: "Pont-Scorff: all the conditions for a ZAD have been met". Of course, this is utter nonsense, since I don't know many landowners who would foment a ZAD on their property. But this is our home, Pont-Scorff, and we're here legally. But the PQR, which is subservient to the local authorities, is trying to demonize us in the eyes of public opinion, to sully our reputation in accordance with the wishes of the local potentates and, by extension, to influence the Prefect's decisions.

The idea is also to ensure a safety net: even if our meeting went well, there could always remain a doubt in the minds of the most impressionable readers as to our dangerousness. And, in fact, the meeting goes rather well. We go along, accompanied by Sea Shepherd's lawyer. A round table is organized to present all the administrations. Jean, the lawyer, then takes the floor:

—Monsieur le préfet, in addition to being a lawyer for the Rewild coalition, I'm also Carlos Gon's lawyer (yes, yes) and a reserve officer, so you'll understand that the mention of a possible ZAD in the columns of *Ouest-France* sounds like a de facto incompatibility.

Jean will quickly reposition the context of the project, all the illegalities noted, the abuse of power by the State services—in short, he sets the record straight. The Prefect remains silent, looks around the table at his interlocutors, composed almost exclusively of government departments, and blurts out:

—Gentlemen, you should know that I know nothing about animals, even though I obviously work with donkeys.

I can't help laughing behind my mask at the look on the veterinary director's face. A long discussion ensues, during which I take apart all the administration's accusations, and we part on a sort of consensus: keep doing your best, we'll find a solution.

We left the prefecture delighted, convinced that we'd changed the paradigm, but as always, there was a grain of sand in the cog. Two days after the meeting, the prefect was summoned to Matignon, leaving the way clear for a successor who would loosen the reins on his administration, which would have a field day.

Contrary to what we might have hoped, the pressure didn't let up over the following months. Strangely enough, we are saved by Covid. The pandemic kept the administration busy enough to let us off the hook a little. Taking advantage of this calmer interlude, we decided to take the initiative once again by organizing a major press conference at which we denounced the smear campaign against us, explaining why, who and what was at stake.

The Covid months will see the start of major works, beginning with the clean-up. The site was virtually abandoned when we arrived, with countless buildings used as storage areas. For this mission, I can count on the invaluable help of my partner, David, who will take the *lead* in these operations which are, to say the least, unappetizing, but so necessary. He'll quickly be integrated into the team, taking it in turns to go wherever he's needed. The advantage of his presence is that he is a talented photographer who will, in addition to everything else, document absolutely everything that has been done, and in particular the life of the zoo, its caretakers and the animals. He was a precious help in this project, often making me feel less alone despite the short time I had to devote to him. Unfortunately, the energy we put into the Rewild project took its toll on our relationship, and I'd like to use this book to pay him the warm tribute he deserves.

Little by little, I realized that in addition to the external battle, an internal one was beginning to take shape insidiously, like a discreet but already perceptible background noise. A year after our installation, the first tensions are appearing, and they don't augur well.

27.
The Wild Dog's Feeding Bottle

Within the first few weeks, some of the coalition's associates jumped ship. Among them were those who had forgotten to tell us they were working with zoos such as Wildlife Angel. The aim of this organization was to train *rangers* in the fight against poaching. However, among its "partners" are Thoiry Zoo and La Flèche Zoo, which are regular contributors to the association's operations[117]. Inevitably, this turn of events didn't sit well with our associate, and risked jeopardizing his "partnership" with the captivity industry. This unhealthy financing guarantees their submission as well as their survival.

As our fund is called "Let's buy a zoo and free the wild animals!"[118], our relationship was bound to weaken, as was the link between Rewild and One Voice, an association against the use of wild animals which relies on French zoos to place animals entrusted to it by the courts... In short, for the good of all, we had to break with certain partners. In fact, Rewild's aim—and I thought I'd made this clear—has always been to stop making pacts

117. https://www.wildlifeangel.org/partenaires
118. https://www.gofundme.com/f/REWILD

with wildlife exploiters, to rewild animals and release as many as possible. Perrine, for Hisa, was pushed out by Lamya, the president of Sea Shepherd, without us realizing at the time that the pretext used was the beginning of her scorched earth strategy. The coalition was quickly reduced to three entities: Biome, the Athénas Center and Sea Shepherd, without the tensions disappearing.

The first grievance was the omnipresence of the head of Sea Shepherd France in the establishment. In theory, nothing serious. After all, thanks to its deep pockets, Sea Shepherd has helped make the Rewild project a reality and injects several tens of thousands of euros every month. However, I'm in charge of a company, and it's becoming less and less tolerable for Sea Shepherd board members to turn up with their kids and dogs, confusing us with a cab, daycare, kennel and private entertainer business for Their Lordships. I have neither the soul of a princely family *boy* nor the time to waste on idle salamalecs.

Other disagreements will become quite clear, for example, on how animals should be managed. Legally, this is my sole responsibility. I am answerable to the law. I'm happy to communicate on the subject with the Rewild board, but when my responsibility is involved, I decide. But more and more things are happening that I don't like at all. For example, Lamya's companion spends whole days stroking a panther he's set his sights on. It's a health and behavior problem. Indeed, when we arrived, we found that the panther was crying, so to speak. She spent her days calling out. We changed her dietary and social regimes, broke her routines, and her wailing stopped. Until the intruder mistook the animal for a big cat. As a result, when the pet owner disappeared in the evening, the calls started up again.

There were also other issues to contend with, such as staff management. Between two *scrapbooking* sessions, Lamya doesn't hesitate to verbally beat up an employee in front of the others, using public humiliation and invective. It becomes difficult to debate, to exchange ideas, to find a way of conciliation, which poses enormous human, relational and technical difficulties. The policy of Sea Shepherd-France is clear and well known to all: "shut up or get out".

As a result, everything is gradually becoming a source of tension. Our duo is on the rocks. Until one day, when Lamya sweet-talks me:

—You know, Jérôme, I've been thinking about it and I think it would be better if you weren't manager anymore.

—I beg your pardon?

—You'd stay on as caretaker, of course, but I have the impression that manager is too much for you. It's obvious you're getting tired.

—What do you mean, tired?

—You're not tired, are you?

What can I say to him? I'm not tired, I'm worn out, worn out, worn out, given that I work eighteen hours a day on a voluntary basis—in fact, I refused to take the salary offered to me until the entire operation was secured, and as long as the employees who chose to stay with us had no guarantee that the structure would actually function (I say this in response to the media and digital slanderers wondering where the money went and implying that I'd made a fortune out of it—not a single euro, kids, not a single one). When I've finished managing the day-to-day, I also have to make sure that our associates are making progress on the project to welcome groups, as we want to turn the ex-zoo into a militant and open space, mixing exhibitions, artistic expression, *vegan food*, yoga and *tutti quanti*, in order to bring in the €50,000 a month our establishment needs to operate. On top of that, we have to deal with media pressure, recurrent summonses to the gendarmerie, and the management of staff, animals and security. If someone who has to put up with this for months on end, without a day off, isn't "tired", he's probably dead!

Especially since money is a constant preoccupation. When I took over the business, it cost €125,000 a month. I'm turning myself into a *cost killer*, but I'm not cutting into the budgets dear to *managers* because I'm hiring and increasing certain employees. On the other hand, I'm turning off unnecessary taps. I discover, for example, that the zoo provides seven or eight cell phone lines, two of which are allocated to former managers who are no longer with the team. Similarly, I find that some suppliers bill us for up to two and a half times the quantity of hay or straw they actually deliver, while pressuring us to settle invoices that predate our arrival. I notice extremely dangerous excesses (particularly with elephants) by an employee whom I have to dismiss, as his unprofessional attitude is endangering his colleagues and Rewild. The dismissal didn't go well, but I had no choice.

For the next year, I became a sort of hypervigilant watchdog, both in material and human terms. Every day, I have two, three, ten new problems to deal with. The death of Jacob the rhinoceros, conflicts with certain former directors and with the children of the former owner, embezzlement in previous financial years, the birth of the wild dogs that the mother has abandoned and that I must bottle-feed, taking care to avoid impregnation so as not to jeopardize an eventual return to the wild... Above all, and as trouble never comes alone, my father is slowly dying of Alzheimer's disease. I'll be burying him on November 1st of this year, in the midst of the Covid restriction.

*

At this point in my story, the reader is entitled to ask again the fateful question: when I recount my troubles, big and small, am I not straying from animal trafficking, which is supposed to be the subject of the book? Not in the least. The fight against animal trafficking doesn't always involve spectacular operations, with adrenaline pumping, large-scale deployment of forces, orange armbands everywhere and emotional successes for wildlife conservationists. It also involves putting in place the conditions necessary to advance the cause, and this is as concrete, down-to-earth, tedious and exhausting as it is indispensable. That's why I think it's also necessary to take a look behind the scenes at Pont-Scorff.

The story of Rewild is the story of a unique experiment in the world, designed to help reduce the illegal keeping of animals. So, even though I'm laminated, stoned, exhausted, no, sorry, Lamya, I'm not leaving for three reasons:

- I'm committed to this project, and I don't know many people who can put up with what I'm going through;
- I believe we're moving in the right direction;
- and I'm sure I belong.

28.
The Pachyderm's Promise

To celebrate the first anniversary of our storming of the Bastille, the caretakers have prepared a big surprise party for me. They know how much energy I expend. They see me toiling away, trying to keep a sense of humor so that none of the staff feel like giving up. Thanks to the sheer mass of everyday crap (yes, the word may shock, but no other would be more accurate), I didn't see it coming, and I'm moved by the warm atmosphere and gratitude of my colleagues. To mark the occasion, David has put together a slide show for me, retracing a year of life on the site, accompanied by dozens of photos of animals and caretakers, and some lovely words of encouragement. I'm celebrating for once, but I don't realize that this touching sequence will seal my break with Sea Shepherd-France, which David didn't mention in his photographic tribute to me. They'll never forgive me for this, as it highlights the over-inflated ego of the organization and its leaders.

For a month and a half, my relationship with Lamya continued to deteriorate. We hardly spoke at all. Volunteers moved into our premises. In mid-February, taking advantage of the fact that I had finally taken two days off, Lamya, assisted by the parasites from Sea Shepherd-France, illegally organized a staff meeting to convince them that I was a danger to Rewild and had to be got rid of as soon as possible. Indeed, I don't like animals, I'm technically very perfectible, I have no sense of social dialogue and, icing on the moldy clafoutis, I try to mask these flaws by manipulating people. The solution is simple: Jérôme Pensu means problems; more Jérôme Pensu means more problems.

Naturally, I hear about this meeting. The next day, as head of the company, I call a counter-meeting, with the participation by video of Gilles and Lorane, members of our board of directors, Lamya and the employees. Courageous, but not too courageous, the vociferous Lamya arrives with her spouse and a whole bunch of volunteers who have absolutely no business being here, but who have come to recite their anti-pensuical text in order to destabilize not only the company, but also the employees. They've seen me work my ass off from morning till night; they know that I've raised their salaries when it was deserved; they've seen that, when it came to fighting for the project, I was always there. And yet, at the end of that meeting, no one speaks to me anymore. Lamya's label of manipulator is the most terrible. Her spell is in full effect. Impossible to argue: that would be manipulation. Impossible to put forward facts: that would be manipulation. Impossible to object to indisputable elements: that would be manipulation. And, conversely, remaining silent is also manipulation!

So I'm taking a step back. I stop the daily *briefings*. I give my instructions over the walkie-talkies. I cut myself off from the others.

*

A week later—need I say that, in my condition, a week is an eternity longer?—Loïck, an animal caretaker, former Sea Shepherd-France volunteer (he took part in several expeditions) and very close friend of Lamya, enters the scene. He can't bring himself to see

Rewild go peanuts. He resigned from the Puy du Fou to work here, and finds himself in a madhouse where a great project is imploding over a silly relationship issue. As a result, he knocks on the door of the large room that serves as my office and asks me some questions. I tell him straight out:

—I'll answer your questions, but please don't take my word for it, and check all my answers.

I show him the shelves around me and add:

—It's all there. From the outset, the documents have been available to anyone who wants to consult them.

After our exchange and some checking, Loïck proposes a four-way meeting: he, Sauveur F., to whom we still owe a good deal of euros, Lamya and myself. I approve the proposal; Sauveur also approves; Lamya, on the other hand, insults Loïck, who goes from being a friend to the ultimate asshole in the blink of an eye. He falls out of the wardrobe when he sees his ex-girlfriend frothing with rage, whereas his idea was to calm things down and avoid going to the wall. The next morning, he turned up at the staff *briefing* and recounted the explanations I'd given him, which he was able to confirm, as well as the hysterical reaction of the chief accuser. For him, the die was cast: if anyone had tried to manipulate the zoo staff, it was Lamya. The motive behind her attempted *putsch*? She was probably hoping to get the zoo back for Sea Shepherd-France by getting rid of the kingpin who was trying to preserve the original spirit of our joint project—me, that is.

*

Two days before the destabilization meeting, the members of the board of directors, including myself, received a letter from Lamya stating that, in view of the sums committed by the NGO, they wanted full powers or threatened to cut off the funds immediately.

They pretend to forget that, while they have put in the dough, no one has twisted their arm, and that others are working, bringing know-how, experience and knowledge to the cause. In fact, the association decided to get its hands on Rewild a long time ago. Case in point: they've created Rewild T-shirts... by adding their logo to the Rewild logo. I suspect this predatory attitude is motivated by internal power plays within their organization. At the time, Paul Watson and Lamya had not yet been fired from Sea Shepherd International. Between the most prominent directors, there was a fierce battle to see who would succeed Paul Watson when he left. Lamya is aiming for the job, but finds herself in competition with some of the organization's captains. Taking over a zoo, which would be the international organization's only base on land, would enable the candidate to hope to establish her authority and set up a care center. Animals are always good for publicity, and publicity is money.

Be that as it may, Rewild's board of directors refused to accept Lamya's exorbitant demands, and the association carried out its threat and cut off our funds. At the same time, it endangers both employees and animals. I was forced to file for bankruptcy. I declare to the commercial court that I'm going out of business. Attempts to evict me redouble. To no avail.

Even when I filed for bankruptcy, I didn't give up. It's a new experience for me and I'm discovering the role of court-appointed administrator. Officially, since filing for bankruptcy, he has taken over the reins of the company, and I've become an executor on behalf of a third party. I no longer have any decision-making power, but I help the administrator to manage the company during the period granted by the court. The court has given us one month to find a way out of this mess and attempt a rescue.

I'm encouraging the caretakers to continue the Rewild initiative by creating a SCOP. Thanks to the support of celebrities, a new fundraising campaign has been launched to try and save the project. It was relayed by Laura Felpain, who had been contacted by a mutual acquaintance. The fund quickly gained momentum, too much for Sea Shepherd's liking. Immediately, a horde of mischief-makers began reporting the situation en masse, alerting the influencer to the fact that, less than three hours after the fund's launch, Hello Asso shut it down and blocked the sums collected.

This new low blow precipitated the company's liquidation, as all our attempts to save the situation were systematically scuttled. At the same time, Sea Shepherd illegally cut off access to our mailboxes and all our communication tools, despite the fact that I was the owner of these as a registrant. Sea Shepherd's treasurer registers the Rewild name with the INPI, preventing us from using it under threat of legal action.

We leave the court: the liquidation has been pronounced with, exceptionally, a month's continuity due to the presence of the animals. I take my things and go to greet the liquidator before leaving, but César, as he's called, insists that I stay for two reasons: firstly, managing a zoo is not in his remit; secondly, he's worried about the bloodshed of the employees, some of whom are quite on edge.

Here I am, in charge of the day-to-day running of the business under the orders of the court-appointed receiver, and still on a voluntary basis. I receive prospective buyers, assist the auctioneer who has come to take stock of the company's assets, and look for funding to ensure that the animals continue to be fed. To my surprise, the Brigitte Bardot Foundation offered me a cheque for €50,000, on condition that I leave. Decidedly! The reason for this codicil is simple: the foundation's director of operations is vice-president of Révolution écologique pour le vivant (REV), Aymeric Caron's political movement... of which Lamya is also vice-president.

Their common goal isn't just to get back at me, even if this very particular spirit is no doubt no stranger to the blackmail carried out by Brigitte's power broker. No, their idea is simply to get me fired from my job so that they can take over the position and better position themselves as the buyer.

To put it plainly, it's very dangerous for Lamya to keep me in my job, since the very person who screwed up the project is in the running to take over the company she sank. The Brigitte Bardot Foundation's method is exactly the same as Sea Shepherd's: if you don't do exactly what I say, you die. On the other hand, if you allow yourself to be bought, you'll be fine. Sinking a company in order to buy it, as others have done before, isn't that the definition of a hostile takeover?

No such luck. For the first time, perhaps, this band of would-be dictators finds itself faced with a lisp that, one, is able to tell them to fuck off, and, two, doesn't give a damn—but with a vengeance!—of the gossip, backbiting and lies that can be spilled over him for weeks on end. Me, I know what I do, what I live for, what I'm worth. The simmering, tenacious, obsessive hatred that Lamya shows me on the networks reaches such a level that it seems almost comical. Unpleasant, perhaps, but almost comical too. On entire digital walls, my ex-partner rants about me, designating me as enemy number one, broadcasting my photo so that everyone knows what the bastard she can't bring to heel looks like.

And Sea Shepherd's *lobbying* finally paid off: one morning, I read in the local press that I had been dismissed from my position as former manager by a decision of the commercial court. As all this jargon is new to me, I don't really understand what it means, and as I'm already in receivership I no longer have any control over the management of the company, so it doesn't really change anything. Well, it doesn't change anything, not really, says a chartered accountant friend of mine, whom I asked for help when I took over the zoo. Under normal circumstances, and in the event of liquidation, the ex-manager retains two rights. One is to have access to all the takeover files and, above all, he's the only one legally able to appeal if the buyer doesn't suit him. End of story, they think. But how boring life would be without its share of little surprises!

*

The next morning, my lawyer in Lorient calls me and asks:

—Are you hanging on tight?

—I have to be, because the way it's been blowing for a while...

—Well, we're going to do a big premiere in France.

—What's that?

—I'm going to summon the public prosecutor to reverse your dismissal.

I approve of his initiative. We win. I'm reinstated as manager. Here I am again, with a front-row seat to witness the violence that continues to rock the administration.

Among the craziest episodes, behind the liquidator's and the prefect's backs, the director of veterinary services attempts to seize animals, notably elephants, in the midst of liquidation, with the aim of sending them to the Tanière-zoo refuge. I learned of this from Loïck, who had his ear to the ground and, quite frankly, marked the vindictive high-ranking civil servant on the knuckles every time he visited. The aforementioned, so sure of his power, took no precautions and, as is often the case with evil-doers, greatly underestimated his adversary, in this case us. The telephone conversation between the representative of the State and those in charge of the Lair is unequivocal: "We've scored a try with today's control, now all we have to do is convert it!"

I now understand why, among the potential buyers I was to receive the previous week, La Tanière was announced, which obviously had no intention of buying, but wanted to come and have a look around. They'd already done the same to Sauveur F. at his zoo. I immediately alerted the liquidator, who flew into a rage and let out a "Oh fuck! He's starting to piss me off! He only just manages to block the seizure". Not enough to discourage the furious DSV who, less than a week later, did it again.

For once, I had decided to arrive at the zoo a little later than usual—barely, in fact. 8.45 am instead of the usual 7 am. As I walk past the visitor parking lot to the service entrance, I see three police trucks, the Psig, the dog handlers. I say to myself: "Fuck, this time it's a seizure".

To recontextualize, we're in the week of my fortieth birthday, when no one in the establishment speaks to me anymore. I open the door to the *briefing* room, where the carers were gathered in conclave, and call out: "Gendarmes in the parking lot". Suddenly, like a flock of sparrows, all the carers run out of the room and disappear, without me even knowing where they're going.

I barely have time to get up to my office, just above the room, before I hear tractors, service cars and pallet trucks arriving. Without my giving a single instruction, they had taken it upon themselves to defend the fortress, forgoing any internal friction in the process. One fetched the welding machine to prevent the service gate from opening, the others metal stakes to be driven into the ground to slow down the passage of hypothetical vehicles coming to fetch animals. One of them blurts out as he arrives with the equipment: "Fuck, I'm going to make a ZAD out of them!"

I immediately call the liquidator once again:

—César? Pensu, tell me, I've got a regiment of CRS in the parking lot. I'm warning you: if you've got it backwards, you'll remember.

—Calm down, Monsieur Pensu," says César, "I'll call the prefect and get back to you."

Not two minutes have passed when he calls me back to tell me that it's just an administrative check. I calm down the caretakers and we welcome the director of the DDPP, his pride and joy, and also on this day, by chance, some of the staff from César's cabinet, who have come to take stock of the company's finances. Today's inspection is purely administrative, says the director. Administrative, I replied. Wasn't the army available?

All morning, the caretakers occupied the office to attend this hostile inspection, to say the least. At the end of the inspection, in a solemn tone, the director of the DDPP said to me:

—I must inform you of the decision I have taken.

—Please do," I said.

—Since you no longer have a manager, and while you're waiting for a buyer, I'm going to propose to the Prefect that he requisition the zoo's former director to ensure the transition. I wanted to ask your opinion.

I think he still regrets it. For the first time, I let loose with the individual. In a tone of voice all my own, and with a vocabulary all my own, I said to him:

—You're never short of imagination when it comes to pissing people off? This zoo has been illegal for 20 years, but you had to hibernate and, all of a sudden, something snapped you out of your torpor and you said to yourself, "I'm going to break someone's balls", and as if on purpose, it happened to be mine. Well, it's not going to be me, because I'm nothing here anymore. The boss is the liquidator, whose representatives you have here today. And the caretakers, good thing they're here too, good luck. And I'll step aside and let them have their say.

What you need to know is that, historically, this zoo has been punctuated by successive directors who have all, in one way or another, vandalized the site for personal gain, either for local glory, or with a view to cheap acquisition, or worse, to misappropriation. And it just so happens that the capacitator proposed by the DDPP is none other than the zoo's former owner, before the founder's daughter thanked him. Suffice to say, the man in question is driven by a sense of revenge. To simplify matters, he is a 30% shareholder in SARL Bretagne zoo. And for the record, the end of his reign was not a happy one, as many of the caretakers remember.

It was Clochette who took the floor, Clochette of course being one of the famous nicknames I've given to the caretakers. She's a young, brittle little woman who nobody expected to play the role of righting wrongs. In a flight punctuated by tears and anger, she pilloried the civil servant, who was now a little less mouthy. The concern in her eyes was all the more evident given that, during the inspection, the caretakers had made friends with the gendarmes, who were obviously less inclined to intervene to calm things down. In the end, he left with his head down, and I escorted the gendarmes who had been watching the control operation back to the gate.

As I greeted the major, I said:

—Major, be cool with my boys. I'm going to leave. They're going to stay. Have you seen what we're asking you to do to them? As far as possible, put your foot down with them...

The major stops dead in his tracks. I freeze too. Let's just say our relations have always been more explosive than cordial. Nevertheless, the officer lets go:

—Monsieur Pensu, rest assured, no one is fooled by what's going on here.

Indeed, I'll realize at the time of the liquidation that the mirage has dissipated, especially when the public prosecutor himself, on the day of the appointment of the buyer, launches into a long *mea culpa* during which Sea Shepherd and the veterinary services take the cake.

*

As we await liquidation, life goes on almost as if nothing had happened, with the atmosphere of a death row inmate who, while awaiting execution, continues to live. We know we've got two or three weeks left, yet we work all day long; then, in the evening, we get together to enjoy the mild May weather in an area set aside as a food court. What's on the agenda? A campfire and a screening of Louis de Funès comedies. We needed to clear our heads and get together again after the big crisis and split that Lamya had tried to create. She, her partner and her entire board of directors have left the premises since the meeting to try and destabilize the company.

The liquidator sent me a list of potential buyers. I received them, showed them around the premises and tried to ascertain their intentions. Among them was a certain Sébastien M. He is a restaurateur from the Costarmorican region, passionate about animals and, by his own admission, has always dreamed of owning a zoo.

He knows how to find the words and attitude that will charm everyone. Conservation, biodiversity, animal welfare and reintroduction are just some of them. Loïck accompanies me to all these meetings, I need to have an additional objective opinion. It's important for the future buyer to have the support of the caretakers, as they could be among those blocking the takeover, and Sébastien M. is well aware of this. He will be visiting the zoo twice, once with his partners, once with his wife. And when we talk to him about the animals, he tells us how much he enjoys getting back to nature, and that one of his major preoccupations will be Gandhi. She's an old female Asian elephant who was captured in the wild 53 years ago and has lived a life of ordeal in this zoo. She has known all the elephants, seen them all die, including the young ones. Her return to a sanctuary in her country of origin before her death would be a fine symbol.

While awaiting the court's decision, and in spite of everything, the caretakers will have supported the SCOP project until the very end, as if, at the last moment, financing could miraculously intervene. In any case, they never stopped believing.

The day of liquidation has finally arrived. It was an opportunity for me to carefully describe to the court the utopian nature of Sea Shepherd's takeover plan, which foresaw an investment of less than €30,000, when the water treatment plant alone required €300,000. The court understood this, and in the end, not surprisingly, it was Sébastien M. who was appointed buyer, the SCOP des soigneurs having withdrawn in his favor. Despite this, I have reservations about the number of visitors he has forecast, which seems to me to be totally utopian. Three years after the takeover, less than 300,000 visitors came to see the "new" zoo, instead of the 500,000 programmed in his takeover project. By 2022, the zoo was running a deficit of almost one million euros.

As soon as he became the boss, the masks fell off and the fine words quickly disappeared. He forbids employees to talk to the former caretakers, and maintains an atmosphere of omertà about everything that happens at the zoo.

To illustrate the method, one evening when an employee was having an aperitif at the house, while talking on the phone with Sébastien M. about her plans to leave the zoo, he concluded the conversation with: "By the way, give my regards to Mr. Pensu"...

How did he know where his interlocutor was? It remains a mystery, but I have to admit that the effect of unease he was looking for was achieved.

As for the new management's program, it's clear and unambiguous. The show parrots, which were free from morning to night when we arrived, are now locked up, the illegal birds denounced in my audit are being illegally evacuated, and the Seychelles tortoises, also illegally held, are being returned to their trafficker owner, despite my repeated warnings to the buyer.

As for the elephants, they never left for Southeast Asia. One of them died within a fortnight, as a result of the changes brought about by the arrival of the buyer and his 25-year-old vet, who knew everything better than anyone else and made such drastic changes to the diet, combined with the change of lifestyle and groomer. The deceased pachyderm's companion, who, out of respect for what these animals represent, should really have been put back in touch with her original ecosystem, was sent to a sanctuary co-financed by the Bardot Foundation in the Limousin region of France. This is a far, far cry from the promises made by the new director, who is filling *Ouest-France* with language that will make you want to smash your skull against bricks, as when he declares: "I want to give to a collective", when what he is doing is emptying the zoo of some sixty "boarders", and not releasing them into the wild.

Under his reign, Rewild has become "Les Terres de Nataé", a zoo that presents more or less legal animals for tickets approaching €25, but claims, like its peers, to "involve the visitor in a strong approach to preservation and animal welfare". He stresses that, like any zoo, it brings in people and therefore contributes to the local economy. No one cares that it profits from wildlife trafficking. Pont-Scorff is back to being just another zoo.

29.
Trunks in Wheels

If I had to sum up the Rewild experience, I'd have to say it was a positive one. We've pulled off a unique feat: buying a zoo from the powerful zoo union, closing it to the public and transforming it into a rewilding center. We've shown the true face of zoos, and brought the issue of seized animals out into the open. We have deconstructed the myths about the longevity of animals in captivity, and their conservation by institutions that are more like gravediggers than charities. In short, we opened a door that will never close again.

The entire zoo industry has mobilized to defend its *business model,* expending considerable energy to bring us down. Their action alone is representative of the visceral fear we've provoked in them. Because, let's be serious for a moment, if we'd been those sweet dreamers, ecologists, anarchists, zadists, as some have described us, we'd have been gently left to crash. Instead, from the outset, the zoos pulled out the sulphate gun, smeared our reputation and mobilized their networks, including government services, to harass us and help weaken the coalition we had formed. I have to admit, it was a good idea. Unity would have been our strength; disunity was our main Achilles' heel.

In spite of everything, and I'm positive about this, Rewild is a great success. Particularly in human terms. The shit we've had to deal with has led to solid friendships. And for that alone, it was worth living. Of course, we would have liked to have gone on from there. However, along with the last few diehards who stuck with the Rewild experiment right to the end, I can claim to have inspired vocations and fanned the flames of anger in some observers who, faced with the scale of the sticks—logs, trunks, forests, even at this stage!—that zoos have thrown in our path, have seen with their own eyes that the aim of these establishments is the commercial exploitation of animals, not their well-being or conservation.

30.
The Impossible Reintroduction of Turtles

The story could end there, but, to give a better picture of the diversity of wildlife trafficking in France, I can't end this book before recounting the big file I discovered in December 2021, which alone illustrates the set of practices that make trafficking possible. When we took over the Pont-Scorff zoo, we had one priority: to make an inventory of the species and individuals held in order to determine which animals would be able to leave, and then set in motion the long and varied procedures required for reintroduction. The list of animals on site includes elephants, giraffes, lions, hippos and one species that particularly caught my attention: Seychelles tortoises, large and beautiful beasts weighing between 180 and 200 kg.

From a technical point of view, bringing these animals back to their natural habitat poses no particular difficulty. Their diet, essentially herbivorous, is very simple. They won't mind eating leaves and grasses in France or the Seychelles. The only thing to watch out for is their reacclimatization. This is obviously taken into account, as it is every time we embark on a process of returning to nature. In reality, the big question for me is who really owns them.

It's a trickier business than it seems. Among the animals we have recovered, some belong to the Pont-Scorff zoo, others have been loaned, others have been donated, and still others have no clear status. First, I discover that the Seychelles tortoises belonged to an Italian zoo. I asked them to bequeath the turtles with a view to their reintroduction. As expected, I received a firm and definitive refusal. Upset, but resigned, I took a closer look at the acquisition documents justifying the animals' presence on the site, in order to have them sent back to Italy in the proper form. And then, surprise! I discovered that the animals had become the property of Pont-Scorff, as evidenced by a certificate of transfer. If confirmed, this discovery would turn the tide, as we would no longer need Italian authorization to send the animals back to the Seychelles.

Going back over the imbroglio, it seems to me that, for some reason, the Italian zoo loaned the tortoises to the Doué-la-Fontaine zoo. Then, through the intermediary of the caretaker who held the *ad hoc* certificate of competence, they arrived in Pont-Scorff, and the said caretaker drew up a certificate of transfer in favor of the zoo. In other words, he gave the animals to Pont-Scorff. How did he manage to do this? A mystery, but that's his problem. It's likely that some complicated, even shady dealings took place, as the caretaker also obtained a tortoise for his own account from the Italian zoo. The main thing is that I have a certificate of transfer.

According to Zim's records, the turtles were born in the Seychelles in 1957. They can be estimated at a minimum of €30,000 each. In other words, the underlying commercial stakes are not neutral. So I applied to the Seychelles government for the animals to be returned to them. My request was accepted. I applied to the French government for the right to exfiltrate the animals. A miracle! For once, the authorities agreed to my request. Everything seems to be going smoothly, but... of course, there's a "but". If there were no "but", there would be no trafficking, it's as silly as that.

*

One day, I see a journalist and a photographer arrive at Rewild, sent by *Le Monde* and validated by Lamya. They want to tell their readers about Rewild. They say they're enthusiastic, but I immediately have my doubts. The questions they ask show that they don't have a high opinion of our work. I've got a feeling these guys are going to do it to us in reverse. This isn't my first day in the world of animal rescue. Over time, my intuition muscle has developed. When I don't feel like a journalist or an observer, what happens next usually proves me right. The interview doesn't go very well. The photographer sneaks some shots and tries to go alone to see the animals. I forbid him to do so, and take the opportunity to crop him.

On January 4, 2021, the article appears and shoots us down in flames. The writer was outraged that we were planning to "release [our] boarders into the wild". And since, against my advice but under the seal of confidence, Lamya has mentioned the forthcoming reintroduction of the turtles into their natural environment, they hasten to reveal the secret project in order to denounce "Rewild's controversial initiative to 'liberate' animals from a zoo"[119]. The Italian zoo immediately sent us a formal notice and took us to the Lorient Commercial Court. Luckily for us, in support of their claims, they made a mistake that should have been fatal. They published a bunch of documents supposedly justifying the legal acquisition of the turtles. However, when I read these documents, I realize that these animals have been trafficked via a money-laundering operation. Sea Shepherd's lawyer, who is supposed to be defending us, pushes us into the ground, perhaps

119. https://www.lemonde.fr/planete/article/2021/01/04/en-bretagne-l-initiative-controversee-de-rewild-pour-liberer-les-animaux-d-un-zoo_6065102_3244.html

out of fear that Rewild will keep its promises to return the turtles to the wild before the NGO takes over. We are condemned to return the turtles to the traffickers.

The decision is absurd and outrageous, but it confirms that laundering illegal animals is even easier than I imagined. Since June 24 1974, despite the many lucrative pressures weighing on this small state, the Seychelles have banned the departure of adult turtles from their island, with two exceptions: for scientific institutions or as diplomatic gifts, and with a limit of 15 individuals per year.[120]

Since then, for these turtles listed in Appendix B of the Washington Convention to be marketed, they must have been born in captivity in a legal, i.e. registered, establishment, and only individuals with a plastron size of less than 33 cm can be marketed. In 2010, the Italian zoo purchased ten turtles from the Seychelles that met this requirement, corresponding to animals between 5 and 7 years old. However, CITES does not require them to be marked before departure. This is the loophole the Italians are going to use to bleach the adult turtles at Pont-Scorff.

The aim of the bleaching operation is to pass off two adult turtles of unknown, i.e. illegal, origin as legally acquired. The laundering phase begins with the arrival of the ten legally acquired turtles. They are identified with transponders. At the same time, the two 1957 turtles are marked in the same way, making them appear to be part of the officially acquired batch. Once identified, they become animals of legal origin, benefiting from the ensuing marketing authorization. What has happened is crystal clear: the zoo has used the importation of young turtles to legalize its fifty-year-old animals.

When I realized this scam, I alerted the public prosecutor. Too late: following the article in *Le Monde*, the reintroduction

120. https://www.sauvage.bzh/sommaire/?file=Chapitre24/loiseychelle.pdf

authorization was withdrawn so that we wouldn't release any animals. As soon as Rewild went into liquidation, the director of the DREAL, whom I had warned of the trickery (which earned me three hours' interrogation at the gendarmerie), sent the turtles back to Italy with the blessing of the buyer whom I had warned. Power to the traffickers and their accomplices!

31.
The Fortune in Corpses

Far from being anecdotal, the failure of the Seychelles turtle reintroduction is highly significant. It shows that a further step has been taken. Traffickers are not only taking advantage of the lack of resources and training of the administration and police. They also benefit from relays in the ranks of those who should be fighting against them. I've noticed that, even when evidence of trafficking and money laundering is brought to the attention of the competent (or supposedly competent) authorities, they don't hesitate to go along with the traffickers, no doubt to avoid alienating the zoological diaspora. In the case in point, the authorities scuppered our project and returned the turtles to a zoo, the source of their trafficking, for three reasons.

The first reason is somewhat personal. If we had successfully carried out the turtle reintroduction operation, it would have proved the validity of Rewild's approach, beyond the economic vicissitudes encountered. The French zoos couldn't accept this. As soon as the Italians initiated proceedings, Céline Serrano, a journalist with France 3, highlighted the fact that the administrative and judicial persecution of which we were victims was no mere epiphenomenon. It was part of a radical determination to put down "an ambitious project, the first in France to propose reintroducing as many animals as possible into their natural environment", which "arouses the hostility of many players in the animal sector, veterinarians, zoo directors..."[121] From this point of view, the turtles are merely the collateral victims of the intense collegiate fury we had to face as soon as Rewild was officially launched.

The second reason is a matter of communication. If, instead of returning them to a zoo, the authorities had seized turtles on the grounds that they had been trafficked, the image of zoos as a whole would have been tarnished. This strong gesture would have revealed that animals held and presented to the public may have been illegally taken from the wild; and this truth would have dealt a blow (temporarily, but a blow nonetheless) to the soothing rhetoric served up year-round by the captivity industry.

The third reason is *business*. Reintroducing animals into their natural environment would have demonstrated that such action is possible, even for individuals who have spent more than a decade in an enclosure. Contrary to what zoos preach, if they don't reintroduce animals into the wild, it's not that their victims can only survive in captivity, it's just that keeping them there earns them millions of euros.

As a result, even when it doesn't concern a French company, even when it doesn't work to blackmail jobs via seasonal workers, the hotel trade and the trickle-down of tourist manna to a territory, the hexagonal zoo lobby can come to the rescue of animal trafficking. Not out of European solidarity with their peers, but to preserve their false image as saviors of animals and biodiversity.

121. https://france3-regions.franceinfo.fr/bretagne/morbihan/retour-aux-seychelles-tortues-geantes-pont-scorff-menace-1911776.html

*

From the above, I draw not a conclusion, but a confirmation: there can be no wildlife trafficking without the complicity of the administration. Consequently, this scourge cannot be defeated without a political, administrative and judicial desire to tackle it. At present, we're a long way from achieving this, to the point where we even seem to be in denial about trafficking. A few outbursts in Qatar, perhaps, in nightclubs, in the odd shady circus... Elsewhere? No, it's long over! Such an assertion is false. Many breeders—especially of exotic birds—are traffickers, and so are many zoos. The administration covers up these shenanigans through:

- incompetence;
- sometimes connivance;
- and often an inability to act due to a lack of suitable structures.

The desire not to see what's going on makes a fine pairing with the protection of the sacred monsters that are zoos... and the professionals involved in animal exploitation.

It's a fact that trafficking doesn't hide. Better still, it no longer hesitates to strut its stuff. The sense of impunity of those involved is unlimited. Why should they? Their industry is flourishing. It's back to where it was in the 1970s. It has no barriers, no modesty and, of course, no remorse. In just a few years, the Gabon Grey, one of the world's best-selling parrots, has found itself classified as a vulnerable species, and therefore on the brink of extinction, putting a definitive end to the myth of sustainable exploitation of biodiversity, which CITES was supposed to guarantee.

Constraints are creeping up on the mass murderers of biodiversity. Quotas? Exceeded. Markings? Falsified. Public display rules? Flouted. Breeding legislation? Disguised. And yet, all is not lost!

Certainly, in France no less than elsewhere, combating wildlife trafficking is not a priority. The seriousness of the situation and its catastrophic consequences for humanity are not taken into account. This is all the more infuriating given that there are practical solutions that are almost simple to implement. I'll close this book by listing them.

32. Six Ways to End Wildlife Trafficking

Environmental Justice

The fight against wildlife trafficking is possible, in fact it's distressingly simple. The only thing missing is the will to do it. Oh, sure, you'd have to put a few tools in place, make a couple of regulatory changes, but on the whole it's easy and inevitably judicial.

Firstly, we need to set up an environmental court that would deal exclusively with cases involving damage to biodiversity. Whether it's a question of environmental destruction, protected river species, pollution or trafficking, this court, composed exclusively of specialized magistrates, is THE solution for real repression commensurate with the crimes committed and the risk suffered by our fellow citizens.

In France, there are specialized courts in every area of life:

- the Social Security Court;
- the industrial tribunal;
- the family court;
- juvenile court;
- the Commercial Court.

There's even a special court for rural leases, but not one for the environment. How crazy! The environment is the issue of the next 50 years. The environment is all the living things that condition our existence and allow us (or not) to survive on this Earth. Important, isn't it?

All cases of environmental destruction, draining of wetlands, river pollution, destruction of protected species and trafficking are systematically dealt with between two cases of burglary, exhibitionism or neighbor disputes. It's true that there are regional hubs specializing in environmental offences. The PREs (for regional environmental poles) were created in 2020, but they are no more than a band-aid on a wooden leg. There is no specialized judge or public prosecutor, which means that environmental law will be applied at *a minimum*.

The jurisdiction of an environmental court should include "the investigation, prosecution, instruction and judgment" of offences under the Environmental Code, the Forestry Code and certain offences under the Rural Code, the Maritime Fishing Code and the Mining Code... In short, a real weapon to crack down on trafficking, entrusted to specialists capable of understanding the stakes, the complexity of the networks and the various operators, following the example of the National Financial Prosecutor's Office.

To achieve this, we need a real political will, an awareness, almost an ecological awakening, but will it happen?

Fighting Corruption

If we want to put a stop to trafficking, we need to start from the long-established facts about its causes, not least of which is corruption, sometimes at the highest level of the environmental police.

It is inconceivable that the former gamekeepers, who have become police officers with the power to detain and search, can only be controlled by the administration. One of the measures to be introduced, at the same time as the environmental court, is the creation of an environmental police force. This would not only put an end to the misuse of procedures, the petty arrangements between friends and the laxity with which some officers, particularly those in the police department, willingly operate, but would also exert a valuable supervisory pressure on their activities. This activity would limit the many abuses faced by environmentalists and, conversely, by traffickers. In other words, the public should get involved...

All the cases I describe in these pages, which highlight conflicts of interest, breaches of probity and corruption, would be totally impossible if there were an environmental police force. Today, any denunciation of these facts arrives directly on the desk of either Gérard B. or the Director of the OFB, both of whom we have seen to what extent they are involved in affairs which, it is not forbidden to hope, will one day end up "blowing up in their faces" as they deserve.

*

Financing the Protection of Biodiversity

The creation of an environmental court and police force cannot be an end in itself. It must be accompanied by other equally dissuasive measures.

These include the creation of a biodiversity tax, which would be an essential funding tool for all the measures to be put in place.

Yes, I know, another tax. But a useful one. If you buy an electrical appliance, you pay a recycling tax that is almost imperceptible in relation to the sale price. Without judging the use of this tax, I don't know the subject, but in any case it was imposed without consultation and obviously without anyone finding anything wrong with it.

Hundreds of companies in France and abroad are making shameless, uncontrolled use of biodiversity.

Importers and users of exotic wood, importers of shellfish, insects, live and dead animals, pet shops, zoos, exotic hide tanneries—all of them have been enriching themselves to the hilt for years without paying their compensation contributions for this disproportionate exploitation. France's 50 biggest zoos alone account for 17 million visitors. A tax of 0.50 euro cents would generate over 8 million euros in tax revenue. Applied to the entire industry exploiting biodiversity, this tax would generate over 50 million euros a year to finance all the essential measures aimed at putting a definitive end to wildlife trafficking.

*

Yes, there is a lot of money to be made. A biodiversity tax is the solution, so we don't have to listen to the government's whining about how there's no money to fight wildlife trafficking.

Except that, at present, biodiversity operators are immune to this kind of proposal, given the blackmail they exert on the State.

Of course, the *lobbies* are against it. They have already obtained a VAT rebate for zoos. In 2014, zoos benefited from a VAT rate of 7%. A plan to increase the latter by a few percent was on the cards at Bercy. Immediately, the princely family in the field pointed out that this would result in the dismissal of 45 people at Beauval alone (and a small cost of blackmail in passing). It was direct pressure on the presidential couple, whom they know well, that enabled them to return to a rate of 5.5%. Yes, you know, the same rate as for basic foodstuffs? Cost of the operation for the State: 7.2 million euros. Imagine what we could do with that kind of money.

I submitted a proposal along these lines to Bruno Le Maire in 2017, when he was Minister of the Economy. Bercy's response: new taxes were out of the question. What a surprise!

And yet, eight new taxes have been introduced under the Macron presidency, proof if any were needed that, where there's a will, there's a way.

Stop Deregulation

At this stage, environmental justice, police control and the financing of the process are all under control. The fact remains that one of the components of trafficking is the misuse of regulations, or the writing of regulations in favor of *lobbies*. This is a very simple matter. We now know all the loopholes, all the fraud techniques and all the tools used to launder money, so we have all the knowledge we need to write a simple, readable text, in line with the international commitments France has long ratified.

To achieve this, we need to re-establish a number of fundamental principles:

- tamper-proof, unalterable and long-lasting identification;
- a ban on the sale, possession and marketing of European fauna;
- establish a list of animals authorized for sale and keeping, with everything else *de facto* prohibited;
- re-establish genetic controls, including administrative controls;
- proof of legality to be provided by the holder;
- put an end to remote declarations;
- eliminate all exemptions from identification requirements from current regulations.

At this point, I don't see how a ministry can justify the existence of such a permissive regulation and refuse to amend it accordingly. If the high-ranking civil servants in charge of drafting it are marked to the hilt, and if real expert opinion is consulted, we should finally have a model of its kind in France, which could also inspire several European countries.

Training Agents

One of the key points in the fight against trafficking is control and the agents who carry it out. Numerous cases have highlighted the enormous shortcomings in this area, even when faced with agents with the best will in the world. Not only must their knowledge of fraud techniques be reinforced, but also their knowledge of the biology of the species encountered, their capture and restraint, and all the keys to determination that will enable them to remove or confirm doubts as to the legality of a detention.

On several occasions, I have been called upon to provide such training, even going so far—and this is a unique case—as to take animals to the OFB training center to introduce them to the capture of large birds of prey or parrots, a subject that is often a stumbling block during an inspection.

This knowledge enables us to counter the arguments put forward by traffickers to evade controls, such as "my animals are in the breeding season", "they'll abandon their eggs", "don't touch the young, they'll abandon them", and so on. The aim is to make environmental police officers specialists in the keeping, fraud techniques and restraint of controlled species.

Handling Seized Animals with Dignity

Finally, the fight against trafficking requires the creation of reception facilities for seized animals. The aim of these establishments would be to return the animals to the wild whenever this option is available, and in most cases, as stipulated by CITES.

Today, it's as if the police, faced with a badly parked car, were to pass by, fatalistic, because they had no operators to remove it and no place to store it. With one difference: we're not talking about objects here, but living beings. Beings torn from their ecosystem. Beings who, at their own level, ensure the ecological balance of the planet. Curiously, however, we can use the automobile metaphor to illustrate the problem. In the absence of an impound lot, let's imagine that some small-time crooks come up with alternatives: one has a plot of land on which to pile up a few cars; another has recovered the use of a parking lot, where he makes a point of collecting the vehicles brought to him in a haphazard fashion. As nothing is done according to the rules, some cars disappear, others come in as Audi Q8s and leave as 4Ls, but with the same reference number, all the vehicles returning to the market without any of them accounting for themselves... By the way, since we're in the car metaphor, would it ever occur to the State to have car salesmen run the impound yards?

When it comes to wildlife control, all things being equal, the analogy is striking. In theory, legislation should put an end to the illegal detention of animals; in practice, it justifies the existence of "shelters" with questionable practices, bogus approvals and disreputable operations.

Against all expectations, the authorities support them, as these places offer a simulacrum of a solution to the few agents who still carry out seizures, for want of anything better. And everything hinges on this "for want of anything better". The "for want of a better word" is not a proof of realism, it's a tremendous support given, consciously or not, to the traffickers and one more ordeal inflicted on the animals, and therefore on biodiversity.

For want of a better solution, we hand over animals to institutions that turn into zoos and zoo breeding grounds.

For want of a better solution, we prolong the detention of animals that should never have been in captivity in the first place.

For want of a better word, we ratify the erosion of biodiversity by acknowledging the French state's inability to fight traffickers.

On the other hand, national refuges delegated to public service, respecting current standards and aspiring to set their residents free, would prevent such abuses and would make it possible—it's an obsession with me, I admit!—to train environmental police officers.

What are we waiting for: an environmental court, a police force, new regulations and a reception facility to serve as a training center, all financed by a biodiversity tax? Soon, most species will be in such danger of extinction that it will be too late, and a few voices will be raised to say: "If only we'd known". Well, I can only say one thing: I know, and now you know too. What happens next will be a demonstration of the will to change, or not.

Postlude

At the end of this book, since I've said "I" throughout, a little personal assessment is in order. Throughout my life, I've never stopped alerting, warning, pleading and raising awareness, always with the same aim: to advance the cause of the environment through that of wild animals. Long before the word became fashionable, I played my role as a whistle-blower with associations, government departments and inspectors, earning me both expert and delinquent status. I didn't care about these differences of opinion. I was and remain driven by a triple obsession:

- to get the message across;
- that the files get out;
- that the bad guys be put out of business.

Today, I can see that wildlife trafficking exists thanks to those who have an interest in covering it up. It's still very much with us, and as difficult as ever to stamp out. In fighting this *business*, I've had to confront more and more often those I thought were brothers in arms, diverted from the fight by status or the lure of a certain celebrity. I didn't do it out of the goodness of my heart, but because they had become part of the problem. So I stand my ground. The tricks, the traps, the slander, the police custody, the tough interrogations—nothing will ever discourage me. Admittedly, after Rewild,

I had a big slump. I wasn't so much crestfallen as appalled at the scale of the problem, which everyone seems to enjoy munching on.

- Zoos create sales, therefore jobs, therefore taxes, which pleases the public authorities, all the more so as these companies excel at concealing their unbridled mercantilism by letting people believe that they are the last bastion against the erosion of biodiversity… when in fact they are making it worse.
- Breeders and certain care centers happily participate in the traffic by feeding zoological ogres with fresh meat, the origin of which is not always legal.
- The big associations are happy to take advantage of the zoological manna to ensure their lifestyle in general and, often, that of their *top management* in particular.
- Despite the good will of some of its officers, the environmental police oscillate between impotence, connivance and support for traffickers.

So, yes, it's true that, faced with this observation, I took a little tap on the back of the neck and thought: "We can't do it, it's lost, they're too strong". Although discouragement is not in my nature, I felt I had reached my limits. Even after what the Rewild adventure had revealed, the hares it had raised, the wolves it had made howl, nothing had moved. Nothing was moving. Nothing would. And yet…

A few years after Rewild came to an end, I said to myself that if voluntary and courageous initiatives were put in place; if, by dint of perseverance, we succeeded in limiting the power of civil servants complicit in illegal practices against wildlife; if we succeeded in mobilizing disinterested volunteers and agents who were both tenacious and committed; in short, if the context suggested the possibility of putting together a structure that would learn from what had happened (and what hadn't!) in Pont-Scorff, perhaps I'd once again have the idea that my convictions, knowledge, know-how and energy could be put to good use. Looking back, I have the impression that I was just a small grain of sand capable of making certain traffickers' well-oiled cogs squeak. On the one hand, the cog wasn't expecting to squeak as I believe I did, along with my associates, collaborators and friends; on the other hand, the squeak it emitted didn't prevent it from continuing to function in the medium term.

As a result, I oscillate between the awareness of a certain failure in the face of the immutable and the conviction that, to a certain extent, I can be proud of what I have achieved or attempted, so that hope is not disqualified. Without ever taking personal advantage of our struggle (on the contrary!), I didn't deviate. I stayed upright. I didn't make deals. I proposed, again and again. I fought. I crashed. Picked myself up. Challenged. I've evolved. Learned. *In the end,* I have no regrets even if, today, I wouldn't do everything the same way.

I'm too familiar with the temptations of certain associations not to be wary of them. Money, of course, but not only and not necessarily. An *ego* to stroke. Valuable company. The glory of media appearances. So many opportunities to be coaxed, circumvented, seduced by our adversaries—in short, let's not beat around the bush, so many temptations to drop one's pants for a little pleasure, a little satisfaction, a little feeling of success for Bibi, not for the cause. This temptation is the Trojan horse of trafficking. There's no better prey for traffickers than personalities who need to be on committees, at symposia, at events where, by rubbing shoulders with so-called "important" people, they feel like they're important themselves... when who cares about their apple? What counts, what should count, is the preservation of biodiversity, nature, the environment, in short, the living without which we'll all die.

I hear, I read, I know that some people gossip, prattle and slander about me. Even if it's not pleasant in itself, I don't mind it for one fundamental reason: at 55, I can say that all the work I've done for wildlife over the last thirty-five good years, I've done voluntarily. I've written some twenty forensic reports, which were worth a lot of money, but for which I never asked the courts for a penny. That was my *guideline,* and I'm keeping it. When it comes to defending wildlife, no one can reproach me for having taken the slightest beginning of euros. Perhaps this helps to fuel the hatred I know I've stirred up. I have the unshakeable claim to have served the general interest and never, ever to have served myself. Not because I'm a saint: because the cause is beautiful, important, essential. It's worth more than me. It's worth more than us as individuals. I imagine that if this conviction animated more of the bigwigs of certain so-called "environmental defence" structures, the fight for the defence of nature and against wildlife trafficking would be much, much, much (I'll stop there) more effective. And that doesn't depend on the state, it doesn't depend on the traffickers, it depends on each and every one of us.

The erosion of biodiversity has consequences that will be fatal for the whole of humanity. Even if the work that remains to be done to turn back the tide is colossal, reversing the trend is still possible. On one condition: that mankind finally decides to care more about its power to live than its purchasing power.

Main Acronyms Quoted

AFdPZ: French association of zoological parks
BEP: brevet d'études professionnelles (vocational training certificate)
BPI: public investment bank
BTSA GNP: brevet de technicien supérieur agricole en gestion et protection de la nature (advanced agricultural technician's diploma in nature management and protection)
CAP: certificat d'aptitude professionnelle (vocational aptitude certificate)
CERFA: official form, compatible with the Centre d'enregistrement et de révision des formulaires administratifs (Centre for the registration and revision of administrative forms)
CERZA: center d'études et de reproduction zoologique augeron (augeron zoological study and reproduction center)
CJIP: judicial public interest agreement
CIC: certificat intracommunautaire
CITES: Convention on International Trade in Endangered Species of Wild Fauna and Flora, otherwise known as the Washington Convention
CNPN: Conseil national de protection de la nature (French national council for the protection of nature)

CRFS: wildlife rehabilitation center
DDCSPP: Direction départementale de la cohésion sociale et de la protection des populations (departmental directorate for social cohesion and population protection)
DDETSPP: Direction départementale de l'emploi, du travail, des solidarités et de la protection des populations (departmental directorate for employment, labor, solidarity and population protection)
DDPP: Direction départementale de la protection des populations (Departmental Directorate for Population Protection)
DREAL: Direction régionale de l'environnement, de l'aménagement et du logement (regional directorate for the environment, development and housing)
DRIEAT: Direction régionale et interdépartementale de l'environnement, de l'aménagement et des transports (regional and interdepartmental directorate for the environment, planning and transport)
EAZA: European Association of Zoos and Aquariums
EBITDA: earnings before interest, taxes, depreciation and amortization
EEE: exotic invasive species
EEP: European breeding program
FDC: French hunting federation
FNE: France nature environnement
IUCN: International Union for Conservation of Nature
LPO: ligue pour la protection des oiseaux (league for the protection of birds)
LVT: loisir, vacances, tourisme (leisure, vacation, tourism)
MMA: mixed martial arts
OCLAESP: office central de lutte contre les atteintes à l'environnement et à la santé publique (central office for environmental and public health offences)
OFB: Office français de la biodiversité (French office for biodiversity)

ONCFS: Office national de la chasse et de la faune sauvage (French national office for hunting and wildlife)
ONF: Office National des Forêts
OVC: objective with constitutional value
PACS: civil solidarity pact
PRE: pôle régional spécialisé en matière d'atteintes environnementales (regional center specializing in environmental offenses)
PQR: regional daily press
RNE: national stranding network
RTMMF: French Mediterranean sea turtle network
SAASS: assistance service for seized wild animals
SNPN: société nationale de protection de la nature (national society for the protection of nature)
UFCS: union française des centers de sauvegarde de la faune sauvage (French union of wildlife rescue centers)
UMSOM: unité mobile de soins pour oiseaux mazoutés (mobile care unit for oiled birds)
WAZA: World Association of Zoos and Aquariums
WPT: World Parrot Trust
WWF: World Wild Fund for Nature

Acknowledgements

A huge thank you to all the agents of the Office National de la Chasse et de la Faune Sauvage (ONCFS), now the Office Français de la Biodiversité (OFB), the Gendarmerie Nationale, the Douane sometimes and, alas, all too rarely the veterinary services, for their involvement in the general interest mission of protecting biodiversity. Through their work, they are the grains of sand that try, day after day and to the best of their ability, to put the brakes on the infernal machine that is wildlife trafficking. I would like to thank them for their trust—sometimes against all odds—and whatever preconceived opinions about me they may have been given.

Thank you to my parents who, despite legitimate concerns, let me go in the directions I had chosen, and who enabled me to develop professionally even though they sometimes found it hard to understand why.

Thank you also, and above all, to all those who, through their attempts to block me, their slander and their treachery, have tried over the years to stop me and prevent me from carrying out my mission. You are undoubtedly the reason for my success. You remain the motivation without which I might have given up long ago.

Finally, thank you to my family and many friends who have supported, encouraged, respected, comforted and, above all, loved me unconditionally. You are the essential batteries of my energy. You believed in me when I was thinking of giving up. You are my lifelong and, I hope, everlasting support.

Table of Contents

Best sellers Max Milo Editions

The Russian Art of War, Jacques Baud

Hitler's banker, Jean-François Bouchard

Confessions of a forger, Éric Piedoie Le Tiec

The Koran and the flesh, Ludovic-Mohamed Zahed

Governing by fake news, Jacques Baud

Governing by chaos, Collectif

A political history of food, Paul Ariès

Mad in U.S.A.: The ravages of the "American model", Michel Desmurget

Mondial soccer club geopolitics, Kévin Veyssière

Putin: Game master?, Jacques Baud

Treatise on the three impostors: Moses, Jesus, Muhammad, The Spirit of Spinoza

TV Lobotomy, Michel Desmurget

www.ingramcontent.com/pod-product-compliance
Lightning Source LLC
Chambersburg PA
CBHW050524240926
51846CB00003B/678
* 9 7 8 2 3 1 5 0 2 2 7 7 9 *